MATHEMATICS

SELF–TAUGHT

THE LÜBSEN METHOD

FOR SELF–INSTRUCTION, AND USE IN THE PROBLEMS

OF PRACTICAL LIFE

I

Arithmetic and Algebra

ADAPTED FROM THE GERMAN OF

H. B. LÜBSEN

BY

HENRY HARRISON SUPLEE, B. Sc.

Member of the American Society of Mechanical Engineers

Member of the Franklin Institute

PHILADELPHIA

WEST CHELTEN AVENUE

1897

GILLESPIE BROS.,
PRINTERS AND BOOK BINDERS,
STAMFORD, CONN.

PREFACE.

The mathematical works of the late Professor Lübsen have for many years enjoyed a wide fame, not only in Germany, but wherever the German language is read and the mathematical sciences are studied. This popularity has been due, not only to the matter contained in the books, but also to the manner of its presentation. The present writer himself had sought, and sought in vain, for books upon Arithmetic, Algebra, Geometry, Trigonometry and the Calculus, which should present these subjects in clear everyday language, free from the technical verbiage in which so many writers love to bury the simplest propositions. Most authors of text books seem to think that it will not do to explain these things in simple language, lest the way be made too easy for the student, or perhaps, lest the importance of the teacher be diminished. Professor Lübsen was not one of this kind, as the following pages will show, and his books were written, as the title pages say: "for self-instruction and for use in the problems of practical life." How well he has succeeded may be inferred from the fact that twenty-seven editions of the present volume have been issued in German; and almost as many more of the succeeding volumes on Geometry, Trigonometry, Calculus and Mechanics have been demanded.

The present writer has had the experience of most professional engineers in studying and applying mathematics, and has found the road often very rough, even with the assistance of able professors, hampered with text books of the orthodox type. Again and again he has been consulted by apprentices, students and mechanics; willing, industrious, anxious to study at home, yet unable to give time or money for professional teaching. These have felt the need of mathematical training. They have found articles in their technical papers which they knew were of utmost value to them, and yet which they could not use simply because of the presence of a few unintelligible mathematical expressions. It is to meet just such demands that the present volume is intended.

The writer's share in the work has been something more than that of mere translation. Many new examples have been added, and some bearing solely upon European practice have been omitted. The method of proving problems by the process of "unitation," due to Mr. Walenn, has been inserted (page 19), and Professor Robinson's excellent approximate method of extracting roots has also been added (page 201).

It is trusted that this effort to place before English-speaking students a treatise on arithmetic and algebra, freed from technicalities and expressed in simple language, will assist many who are compelled to study alone; and perhaps it may not be found unwelcome to some who wish to "brush up" their rusty mathematical information, acquired long ago under scholastic methods, and not improved by years of disuse.

HENRY HARRISON SUPLEE.

January, 1897.

ARITHMETIC AND ALGEBRA.

First Principles.

If it were possible to produce an introduction to the science of mathematics which should give a broad, comprehensive view and insight into this marvellous creation of the human mind, such a treatise could not fail to produce the most inspiriting effect upon the emotions of the student.

But the innumerable variations and conditions which accompany all the details of this branch of knowledge make such a general view a practical impossibility. It is necessary to assume a certain amount of preliminary knowledge on the part of the beginner, in order to make even the simplest introduction intelligible. Without a knowledge of the system of numeration no one can learn addition; without these, multiplication cannot be understood; while all are necessary for the study of division. So it goes on throughout the whole range of mathematics; the principles overlap and lead into each other. From the simplest propositions we advance step by step upwards, and the higher branches are accessible to all who have climbed the lower steps. No principle can be properly understood if one has not fully mastered everything which has preceded it and upon which each fact hangs like a link in the chain of which it forms a part. In fact the word Mathematics belongs not to one, but to several branches of knowledge, (Arithmetic, Geometry, Mechanics, Optics, Astronomy, etc., etc.) each of which is daily extending its scope, and each of which is practically without limits.

A true conception of the meaning of the term Mathematics, both in the pure and the applied branches of the science, can only be obtained by continuous and earnest study, but we can assure the student that every hour spent in the study possesses both a theoretical and practical value.

Before entering upon the immediate subject, the following explanations will serve to give a notion of the first principles upon which all the others are founded.

I. No thing is *of itself* either large or small; its magnitude can only be spoken of in comparison with some other thing of the same kind. If, for example, one speaks of a large house, the speaker really has another house in mind with which he mentally compares it, even if he does not say so. Such a comparison, however, gives no clear idea as to *how much* larger the one house is, or how much smaller the other one. In order to know this we must have some *unit* of comparison, and consider how many times larger than this unit is the object we have under consideration. If, for example, one has a clear idea of the length of a yardstick, and is told how many yards distant one point is from another, he will without seeing them be able to form an idea of the distance between the two points. This comparison of a size with the mental vision of a unit holds true for all magnitudes, such as length, surface, force, time, etc. It is interesting to notice how meaningless to some people statements will appear which are perfectly clear to others; a farmer knows how big a ten-acre lot is, a sailor estimates that a ship is going at the rate of 15 knots, a machinist grasps all dimensions in inches and fractions thereof, while the dressmaker knows no unit smaller than a yard.

II. All magnitudes have this feature in common, namely, that they consist of parts. They are divided into two kinds, according as the successive parts are considered as absolute groups or as parts of a continuous whole, and hence are classified as *constant* or as *variable* magnitudes.

A number, for example, is a constant quantity, consisting as it does of a collection of non-coherent units. As one of the old mathematicians put it: "One cannot coil pease like rope."

Variable magnitudes are those which possess a continuity, such as length, surface, time. For example, one can always conceive of a line as composed of a succession of smaller lines so combined that end of one coincides with the beginning of the next, all thus forming an unbroken or continuous whole.

III. Every quantity can be only compared with and measured by an article of a similar kind. Beginners must be especially careful to observe this point. One cannot, for instance, *weigh* time. The measure of time must itself be a quantity of time, the measure of surface a surface, the unit of angular measure must be an angle, etc. etc.

IV. A unit may be subdivided into smaller values, or conversely a number of units may be combined to form a larger unit. This fact is seen in all the various kinds of measurements. For example, in units of length we have feet and inches, meters and centimeters; in units of weight, ounces, pounds, tons, grammes and kilogrammes; in units of time, hours, days and years. In mathematics every magnitude practically may become a unit and is so called when it is used as a standard for comparison of magnitude.

V. A *multitude* of any article is therefore indeterminate, but a *number* becomes a determinate quantity of the same thing, or a statement of the relation which that quantity bears to the unit or quantity by which they are to be measured. The *name* which is given to the magnitude indicated by the number is called the *numeral.* This name also indicates *how many times* the unit exists in the number so named, and hence for every number there is a distinctive and separate name.

It is not sufficient that each name should distinguish the number to which it belongs, it must also indicate the *order* in which the number appears with regard to the other numbers; a numeral, therefore, indicates not only a number but also the next preceding number, etc. For example, one cannot obtain a clear notion of the number five without also realizing that the preceding number is four, &c., back to the beginning. It would be quite impracticable to learn to count in any irregular fashion, such as : five, three, one hundred, two, &c.

> "If first and second ne'er had been
> There could no third and fourth be seen."

VI. The act of repeating the numerals in regular order, that is, increasing uniformly, each being greater in value by a unit than its predecessor, is called *counting* If the unit upon which the series of numbers is based has a known value, the value of any of the numerals is also known. If, however, the unit merely indicated that *any* article, of *any* value, is taken once, it and all the numbers based upon it are *abstract* numbers of unknown value. For example : five feet is a quantity of determinate value, because we know the value of a foot, while the number five alone is an *abstract number.*

That portion of mathematics which treats of the derivations and combinations of numerical values is called Arithmetic.

Its fundamental conceptions are : Magnitude, Unity, Manifoldness. These conceptions, which are so simple as to be understood by every one, and which therefore cannot be defined more clearly than by simply naming them, form the foundation upon which the entire science of mathematics is based.

The perception of magnitude is one which must be experienced by every one, and since it is inseparably united with the notion of number it follows that the formation of numbers is almost the first effort of infant intelligence. Almost the first thing which a child perceives is that "many" is composed of "one" and "one" and "one," &c., and thus the notion of "manifoldness" is grasped almost naturally, without effort or instruction.

The study of the formation of primitive and arbitrary names for numbers, varying as they do in different languages, belongs properly to the science of language ; yet there is a certain degree of mathematical knowledge involved in the subject, since the words are so constructed as to assist the memory in retaining them. The formation of the names and symbols by which numbers are known may therefore properly be considered the first subject and exercise in arithmetic. In order then that all which follows may be as clear and easy as possible, we shall take nothing for granted, but assume that the simplest rudiments of arithmetic are to be explained, beginning with the origin of numbers, and gradually proceed to open door after door and penetrate into the innermost and highest chambers of this temple of knowledge.

PART I.

Special Arithmetic.

BOOK I.

Numbers and Number Systems.

1.

The idea of unity or the consideration of a thing singly is expressed by the word *one*. The repetition of unity, however, may be expressed by some other arbitrary word, so that instead of saying one and one, we use the shorter expression *two*, (or in other languages, *zwei, duo, deux,* etc.) For the number following two, we say *three*, instead of two and one; and going thus forward step by step we say *four* instead of three and one, *five* instead of four and one, *six, seven, eight,* — — —, each new name expressing a number greater by unity than the one which preceded it. By constant use and force of habit the mind becomes so accustomed to the order and value of these names that the number of units expressed by each is instantly remembered.

It is not difficult to proceed for some little distance in this way, nor to find names for numbers, such as nine, ten, eleven, twelve, etc., but to carry any system of arbitrary names very far would be a practical impossibility, involving a capacity for inventing names and a memory for retaining them which no man possesses.

Unless, therefore, as Josephus reports, Adam was created a natural mathematician, the world probably waited a long time for some one to invent the simpler method of making names for numbers by using names only as far as ten and then repeating these in combination with the words "ten and." Thus instead of choosing a new name for the next number to ten, might be said : one and ten : then, two and ten : three and ten, &c,, up to ten and ten. The desire for abbreviation naturally soon caused the dropping of the conjunction "and", and the compound words soon became contracted into one-ten, two-ten, three-ten (or in the Latin *undecem, duodecem*), from which our further numbers are derived, e. g. thirteen, fourteen, fifteen, &c. The attempted improvement by giving arbitrary names to the numbers eleven and twelve hardly seems to be any advantage.

The number ten may also itself be used as a simple unit, and (as already explained in No. IV of the introduction) become the base of a new system ; using "twice ten," instead of ten and ten, and proceeding : one and twice ten ; two and twice ten ; three and twice ten, &c. The same idea extends to the series : thrice-ten ; one and thrice-ten ; two and thrice-ten ; three and thrice-ten ; &c., proceeding thus to ten times ten. By contraction again the words : twice-ten, thrice-ten, &c., become : twenty, thirty, forty, &c., except that for : ten times ten, the arbitrary word : *hundred*, is used.

As we advance to numbers higher than a hundred we use this word followed by the smaller numbers, thus : hundred and one, hundred and two, &c., until a second hundred is reached, when we say two hundred, instead of hundred and hundred. We then proceed : two hundred and one, two hundred and two, &c., &c., and so go on till we reach ten hundred, for which the new word *thousand* is introduced. This number thousand is then used as a new base through the quantities ten thousand, and hundred thousand, neither of these having special names given to them. For a thousand thousand, the word *million* is used, and for a thousand millions, the word *billion*. We may thus proceed indefinitely with a million billions, or trillion ; a million trillions, or quadrillion, and so on, but these are quite unnecessary, even a billion being rarely used except in indicating the distance of the fixed stars, or similar astronomical magnitudes. See Appendix, § 311.

2.

In the manner above explained it is therefore possible to name all conceivable numbers by the use of very few words.

Starting with the names of the first ten numbers, and adding to these the words "hundred," "thousand," "million," and at the most, "billion," we can with fourteen words provide names for all numbers with which we shall have to work. We are able to count by having memorized these names in childhood almost in play.

If, therefore, we have the fundamental principles clearly grasped, we shall be able to understand the meanings and values of the successive numbers and proceed, not in a hap-hazard manner but with orderly, systematic method, and learn without serious trouble to understand the more difficult portions of our subject.

3.

Although numbers have thus been given names, yet it is evident that without some simpler method of indicating them it would be very difficult, indeed almost impossible, to make calculations. For a long time our forefathers practiced a sort of laborious art of mental calculation (in Europe even as late as the 14th century), until some inventive genius conceived the idea of indicating numbers by simple symbols. Hence we have the symbol 1 to indicate unity, the mark 2 to act as a mnemonic sign for the word two, and so the symbols 3, 4, 5, 6, 7, 8, 9, to stand for the numbers from three to nine.

Originally these numbers must have been quite arbitrary; it was only after continued modifications that they reached their present forms. The invention of these figure symbols has been traced to India. From thence they were acquired by the Arabs, who endeavored to keep them a secret. Only from the time of the Crusades have these symbols been known in Europe, the Europeans calling them Arabic numbers, although the Arabs themselves properly named them Indian. Although isolated examples of their use in inscriptions have been found at earlier dates, the intelligent use of these numbers was even yet scarcely known in the 15th century.

4.

In the same manner as for the numbers from one to nine it would be easy to extend the use of single symbols for the succeeding numbers; for example, for ten we might use the sign

†, for eleven the sign *, etc. But the application of inventive genius has devised a vast improvement over such an unlimited extension of arbitrary symbols.

Suppose we extend the principle adopted in indicating the first ten numbers by taking this group of ten numbers to form a new unit, and in order to distinguish it from the simple units, we will call it the unit of the *first* rank, or "tens." Proceeding thus until we reach ten units of the first rank, we obtain ten "tens," which we then take as a new unit of the *second* rank, or "hundreds," and so continue, making ten units of any rank constitute one unit of the next higher rank. Thus "thousands" would be the *third* rank, "ten thousands" the *fourth* rank, "hundreds of thousands" the fifth rank, "millions" the sixth rank, etc. By thus using a system of successive ranks we may indicate all numbers by combinations of the nine original, arbitrary symbols.

Each of the nine figures 1, 2, 3, 4, 5, 6, 7, 8, 9, when standing alone, signifies the number of units originally attributed to it, but figures, when used in connection with each other, have also an additional meaning according to their positions, with regard to the foregoing system.

In this way the eye can recognize instantly which rank of units is indicated, whether hundreds, or thousands, etc. If, for example, we see the number 6 placed in the third rank, we know at once that 6 units of the 3rd rank are meant, corresponding to the number six thousand. In order to write the number thirty thousand with figures, the figure 3 is written in the position of the *fourth* rank, because thirty thousand is equal to 3 units of the fourth rank. When a number is composed of various units, each one is placed in its corresponding position. For example, the number thirty-four thousand two hundred and fifty-six is separated into its parts, thus : thirty thousand, four thousand, two hundred, fifty and six, and each placed in its proper rank, as shown in the following diagram on line (*a*).

.	.	Rank V. Hundreds of Thousands	Rank IV. Tens of Thousands	Rank III. Thousands	Rank II. Hundreds	Rank I. Tens	Units
(a)			3	4	2	5	6
(b)		7	...	9	...	...	5

In like manner the number seven hundred and nine thousand aud five is written as shown on line (*b*), in which the places of the 1st, 2nd and 4th rank are left vacant.

If there are no vacant places between the various figures, as shown on line (a) the rank of any figure is at once indicated by the number of figures which are on its right; the figure 5 on line (a) is of the first rank, or tens, because there is one figure standing on its right. In such an instance, where all the places are filled, the diagram may be omitted and the number on line (a) written directly 34256, since a little practice enables the rank of each figure to be determined at a glance, by constantly keeping in mind the names units, tens, hundreds, thousands, etc., reading from right to left. When, as on line (*b*), all the places are not occupied by figures, the omission of the diagram would render it impossible to determine the rank of the figures unless some mark or symbol, without value as a numeral, be chosen to fill the places not occupied by numbers. The symbol universally used for this purpose is o, called in English "naught," "zero," "cipher," (in Arabic it is named Zephirum, Zero, Ziffer.)* Beginners very often fail to obtain a correct idea of the meaning of this symbol. As here explained, it only gives rank and meaning of the numbers which precede it, and is without the slightest influence on those which follow it.

5.

We have now no difficulty in naming any number which we may see written in figures. It is only necessary to remember that the thousands, tens of thousands and hundreds of thousands are named in one group, the hundreds are named next, and the tens and units grouped together. Thus 326,000, is called, three-hundred-and-twenty-six thousand; not three hundred thousand, and twenty thousand, and six thousand.

The following examples will serve to show how written numbers are named:

51207, fifty-one thousand two hundred and seven

509004, five hundred and nine thousand and four

319039, three hundred and nineteen thousand and thirty-nine

*Of theee names it is preferable to use zero, not only as understood in many languages but also to avoid the not uncommon blunder of saying "aught" for "naught."

100070, one hundred thousand and seventy

111111, one hundred and eleven thousand one hundred and eleven

555555, five hundred and fifty-five thousand five hundred and fifty-five •

700009, seven hundred thousand and nine

If a number consists of more than six figures, it is divided into groups consisting of three figures, except that the groups at the extreme left may contain fewer than three, according to the size of the number.

Example :

Billions	Millions	Thousands	Hundreds
975	403	507	001

which reads : nine hundred and seventy-five billion four hundred and three million five hundred and seven thousand and one.

If it is desired to write a number in figures as it is read in words the places may first be indicated by groups of points and then the figures written in afterwards until the student can perform the operation with facility without such assistance. Thus for the above number :

Billions	Millions	Thousands	Hundreds
. . .	. . .	. . .	. . .

6.

The above described method or system by means of which all numbers may be written by the use of very few symbols is called the system of *numeration*. In this simple and apparently universal system the fundamental principle needs only to be memorized, namely : that each unit of any figure is *ten times* greater in value than a unit of the figure standing at its *right*, so that we say that our system is based upon the number ten, and call it the Decimal System, (from the Latin *decem*, ten). See Appendix, § 312.

BOOK II.

The First Four Rules.

1. Addition.

7.

The first application of the number system which we have now acquired, will be the method of uniting several numbers together so as to make a single number of them.

The number which we thus obtain by uniting several others, and which is as great as all of them taken together (that is, contains the same number of units), is called the *Sum*, and the method of obtaining this sum in a shorter way than by counting the units one at a time, is called *Addition*. To perform addition it is necessary to assume that the sum of any two single numbers is already known. Addition then involves no very great degree of inventive ability, since the principle involved necessarily follows from the system of numeration.

The numbers which are to be added together are written under each other in such a manner that numbers of the same rank stand directly under each other; units under units, tens under tens, hundreds under hundreds, etc. These columns are then added successively, beginning at the bottom of the units column, and carrying one unit to the next column for every ten units obtained by the addition of the figures in any column. Thus the number of tens obtained by adding up the units column, is carried on to the tens column, the number of tens in the tens column carried to the hundreds column, etc. If there are very many numbers to be added together it may be found less fatiguing to divide them into groups, adding each group separately and then adding the sums of the various groups together.

It is evidently a matter of no consequence in what order the numbers are taken, as the sum must obviously be the same in . any case.

The following examples will be sufficient to show the process :

$$
\begin{array}{r}
789959 \\
98879 \\
357768 \\
5599. \\
99075 \\
800 \\
997997 \\
60088 \\
7673099 \\
\hline
10083264
\end{array}
$$

			7	0	
	4	2	7	5	
		5	9	9	
			9	0	
7	8	4	0	7	
			3	4	
4	2	5	4	9	
1	2	6	0	2	4

Facility in addition comes only by practice, and accountants, salesmen and others who are constantly adding long columns of figures frequently add two, three or even five columns simultaneously, thus greatly increasing the rapidity of the work. Such feats, however, do not imply that the individual possesses any superior mathematical talent in other branches of the science.

2. Subtraction.

8.

In addition we have seen how to find the sum of two or more numbers. Suppose now, that we have two numbers of which we know the sum; which sum, for example, is 12; and also that we have given one of the two numbers, for example, 7. The operation of finding the other 5, by taking the given one from the known sum, is called *Subtraction*. The given number 12 from which the other given number 7 is taken, is called the *Minuend*, and the latter number 7 which is subtracted, is called the *Subtrahend*. The number 5, which is obtained as the result of the subtraction, is called the *Difference*, or *Remainder*. The rule of subtraction is really contained in the theory of the number system and in addition, and the process is just the reverse of addition. The subtrahend is placed under the minuend in such a manner that units of the same rank shall be under each other, and then beginning at the units place each figure of the subtrahend is subtracted from the figure of the minuend which stands directly over it. If the figure in any place of the minuend is smaller than the one which is to be subtracted from it, the figure standing next to the left in the minuend is diminished by one unit and its neighbor to the right increased by ten units, which will enable the subtraction to be made. If the figure next to the left is a zero, we must go on to the next figure to the left and take one unit from it; then call the zero equal to 10, and take one of these units to increase the desired figure by 10. This will leave a value of 9 where the zero stood, which must be considered in the subtraction of the figure under it. This operation will be more readily understood if after any subtraction the remainder be added to the subtrahend, and as the sum should just equal the minuend this will also serve as a proof of the correctness of the work.

Examples :

(1)	Minuend	789		(2)	3453
	Subtrahend	246			1914
	Remainder	543			1539

(3)	30002503		(4)	70040321
	27494097			29067332
	2508406			40972989

3. Multiplication.

9.

If we have to add several *like numbers* together, as for example :

$$
\begin{array}{r}
7 \\
7 \\
7 \\
7 \\
\hline
28
\end{array}
$$

we may simplify such a case of addition by means of *Multiplication* so as only to require two numbers.

One of these numbers is the one which is used repeatedly, (in the above example 7) and is called the *Multiplicand*; the other is that which gives the *number of times* the addition is performed, (in the above example 4), and is called the *Multiplier*.

Very often no attempt is made to distinguish the two numbers by different names, but both are called *Factors*. If therefore we desire to indicate the sum of four sevens, for example, we may do so by the following arrangement:

$$\text{Multiplicand} \quad 7 \;\Big\}\; \text{Factors}$$
$$\text{Multiplier} \quad\;\; 4$$

$$\text{Product} \quad 28$$

The manner in which the important and valuable process of multiplying together two factors each consisting of several places of figures was invented, may perhaps best be made clear in the following manner.

Suppose, for instance, it is required to find the number which corresponds to 4253 times the number 8067, and that so far we only know how to multiply single figures together. We should doubtless proceed somewhat in the following fashion. Considering the example as a problem in addition we think at once of taking the multiplicand, and writing it again and again in a vertical column as many times as there are units in the multiplier, and finally adding all these together. But this is a very tedious operation, and in order to shorten it, we examine the multiplier, and the idea occurs to us to separate it into its ranks as follows:

We can see here that there are 3 units, 5 tens, 2 hundreds and 4 thousands, so that if we divide up the operation of addition into four sums, and take our multiplication as many times as there are units in each rank we have shortened the work very much, as follows:

$$
\begin{array}{r}
3 \\
10 \\
10 \\
10 \\
10 \\
10 \\
100 \\
100 \\
1000 \\
1000 \\
1000 \\
1000 \\
\hline
4253
\end{array}
$$

a	b	c	d	e
	80670			
	80670		8067000	24201
8067	80670		8067000	403350
8067	80670	806700	8067000	1613400
8067	80670	806700	8067000	32268000
24201	403350	1613400	32268000	34308951

Here it will be seen that we have made use of our knowledge that a number is multiplied by 10 or 100 or 1000 simply by placing the corresponding number of zeros to the right of the multiplicand, and so have greatly simplified the work. We have in fact found the result of multiplying 8067 by 4253 by adding the multiplicand to itself as many times as the multiplier has units, as at a, then again as often as the multiplier has tens, as at b and placing a zero at the right, etc., as c and d, and finally adding all these sums together as at e.

Now we see that the largest figure in the multiplier in any case cannot be greater than 9, hence it follows that in this method the multiplicand can never be required to be added to itself more than 9 times.

If, therefore, we make for ourselves a little table showing the products of all numbers from 1 to 9 we shall shorten the work still more by getting rid of all the preliminary additions a, b, c, d.

MULTIPLICATION TABLE.

1	2	3	4	5	6	7	8	9
2	4	6	8	10	12	14	16	18
3	6	9	12	15	18	21	24	27
4	8	12	16	20	24	28	32	36
5	10	15	20	25	30	35	40	45
6	12	18	24	30	36	42	48	54
7	14	21	28	35	42	49	56	63
8	16	24	32	40	48	56	64	72
9	18	27	36	45	54	63	72	81

This should be committed very thoroughly to memory, after which any example in multiplication can be performed as follows :

```
Multiplicand    8067  ⎱
Multiplier      4253  ⎰ Factors
                ─────
               24201
               40335
               16134
               32268
               ─────────
Product     34308951
```

Here we have simply multiplied by each figure of the multiplier successively, placing each product one place further to the right than the preceding, and the result is the same as in the method first shown.

10.

If either or both of the factors terminate with zero, it is only necessary to multiply the significant figures together and annex to the product as many zeros as are at the end of both factors, as shown in examples (1) and (2) below. If zeros occur in the middle of the multiplier the next product is moved as many places to the left as there are zeros, and in order to avoid errors, dots may be used to mark off the places, as in example (3), below.

(1)	(2)	(3)
32	5302000	30794
4000	3400	200506
─────	─────	─────
128000	21208	184764
	15906	153970 ·
	─────	61588 ··
	18026800000	─────
		6174381764

The relative positions of the factors are interchangeable, and it is generally simpler to take the smaller of the two for the multiplier. See Appendix § 313.

In performing multiplications of large numbers together, especially when several large factors are to be multiplied, there is liability to error, either by neglecting to carry the proper amount from one column to the next, or in some trifling detail in the addition. In order to check up the work and prove the correctness of such multiplication, the method of proof by *unitation* will be found most useful.

Unitation consists in reducing any number which is composed of more than one figure, to a number of a single figure, by *adding together* all the figures of which it is composed. Thus the unitate of 2132 is equal to $2+1+3+2=8$; the unitate of 513 is $5+1+3=9$, etc.

If the sum of the figures in any number is a number of more than one figure, then these are to be added together also, until a single figure is obtained. Thus the unitate of 2897 is $2+8+9+7$ $=26=2+6=8$; or of 978596 is $9+7+8+5+9+6=44=4+4=8$, etc., etc. This being understood, we have the following simple method of proof for multiplication. If we have two factors multiplied together, we also take the *unitates* of the two factors and multiply *them* together. (If the product comes to more than one figure, add these together, forming *its* unitate.) The number so found will be equal to the unitate of the product, if the multiplication has been correctly performed. If it is not so, then there must be some error.

In other words: *The product of the unitates of the factors is equal to the unitate of the product.*

This rule holds good for any number of factors, and may be applied successively as the work is performed, or only to the final product, as may be desired.

A few examples will make the operation clear.

Unitate

$532 = 5+3+2=10=1+0=1$
$876 = 8+7+6=21=2+1=3$

3192	Product of unitates=
3724	$1 \times 3 = 3$
4256	

$466032 = 4+6+6+0+3+2=21=2+1=3$

Unitates

82543	=	4
2541	=	3

82543
330172
412715
165086

$4\times3=12=1+2=3$

209741763	=	3
7254	=	9

838967052
1048708815
419483526
1468192341

$3\times9=27=9$

1521466748802 $=54=9$

In the second example we find that the unitates (4 and 3) of the first two factors, multiplied together give $4\times3=12=3$, and that this is the unitate of the product also ; showing the work to be correct thus far. We also have the unitate of the third factor to be 9, and hence the unitate of the final product must be $3\times9=27=9$, which is the case.

We might, however, have deferred the proof until the multiplications were all performed and then multiplied the unitates of all three factors together to obtain the unitate of the final product, thus—

$$4\times3\times9=108=9$$

and then if the unitate of the product should come out $=9$, the work would be proved, if not, some error must have been made.

It is barely possible that *two* equal errors might have been made in separate columns by which the actual figures in the product would give the correct unitate, and not reveal the errors, but the probabilities against such a coincidence are very great.

If the unitate of the product comes out correct, it is therefore a very strong presumption in favor of the correctness of the work, while if the unitate does *not* come out correct, there has *certainly* some error been made.

4. Division.

11.

...ction is related to addition so we have related to
...n another rule, i. e. *Division.* By means of multipli-
...may obtain the desired product of any two given num-
...r example the product 54 of the two factors 6 and 9.
... reverse the operation, having given the product 54,
of the factors, say 6, we find by division the other factor
...is rule gives the answer, then, to two questions; first,
...reat is the sixth part of 54? and second, how often is 6
...ned in 54?

...both cases we are seeking a number (9) which multiplied
the given factor (6) produces the given product (54). This
...ter quantity, the product of which we seek one of the factors,
called the Dividend; the given factor is called the Divisor, and
...he factor which is sought is called the Quotient. In performing
...he operation we say:

54 divided by 6 gives 9, or: 6 is contained in 54, 9 times.
Here 54 is the dividend, 6 the divisor and 9 the quotient.

If a number of several places is to be divided by another
number of several places, as for example, 34308951 divided by
4253, we find the method by the following considerations. We
might subtract the latter from the former number repeatedly,
placing a unit in the quotient for each subtraction, but it is evident that this tedious process may be much abridged by the use
of multiplication. If we compare the number of figures in the
divisor with those in the dividend we see at once, by multiplying the divisor (mentally)by 8 and annexing three zeros, that
8000 times the divisor is smaller, and 9000 times is larger, than
the dividend. The quotient must lie between 8000 and 9000,
and 8 must be.first figure of the result. We may therefore subtract 8000 times the divisor at once. If now we again compare
the divisor with the remainder, we see in the same manner that
60 times the divisor is smaller, and 70 times is greater, than the
remainder, and that hence 0 and 6 must be the second and third
figures of the quotient, etc.

Divisor	Dividend	Quotient
4253	34308951	8000
	34024000	60
		7
	284951	
	255180	8067
	29771	
	29771	

or more briefly :

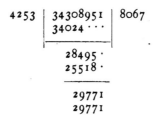

4253	34308951	8067
	34024 · · ·	
	28495 ·	
	25518 ·	
	29771	
	29771	

12.

From the four rules of Arithmetic already explained (in which we see multiplication to be a shortened form of addition, subtraction a reversal of addition and division a reversal of multiplication) we may obtain three additional rules; powers, an extension of multiplication, extraction of roots, a reversal of powers, and logarithms. In elementary arithmetic, the first four rules only are necessary.

The ease and rapidity with which these may be used follow naturally from the simple arrangement of our number system, in which the value for a unit of any figure is a constant multiple of the unit for the same figure standing in the next place to its right, rendering it unnecessary to think of the *name* of the unit at all in practice. If instead our present system we had one in which, for example, six units of the right hand place made one of the second rank, then four of the second made one of the third rank, etc., etc., what would become of our simple and easy arithmetic?

BOOK III.

Symbols, Special Terms, Properties of Numbers, Division, Etc.

13.

Every branch of knowledge has its own special terms and symbols, the use of which assists very much in shortening explanations and simplifying expressions.

In order to avoid subsequent misunderstandings it is important that the meanings and uses of these special terms should be clearly understood at the start and learned thoroughly, so as to be remembered thereafter.

The use of such special terms in mathematics will readily be admitted, and as an illustration of the simplification of a sentence which is obtained by a few simple terms, the following will serve :

"The *remainder* added to the *subtrahend* gives the *minuend* again." If we attempt to make this simple statement without using the special terms subtrahend, minuend, etc., we find ourselves obliged to use the very roundabout expression :

"The number which has been taken from another number, added to the number which remains, will give the first number again."

Like other scientific terms, the words used to denote special terms in mathematics are nearly all taken from classical sources so that they are practically the same in all modern languages.*

It is very desirable that the correct mathematical terms should be used from the very beginning, as even small children would thus learn to associate the words with the *ideas* and have no further trouble.

---

*It is a matter for regret, that the present tendency in Germany is to form scientific terms from German roots, and thus place Germany outside of the international usage in this respect.—H. H. S.

14.

In order to indicate briefly that two or more quantities are to be added together the sign of addition, *plus* (+) is placed between them, it being of no consequence as to the order in which the quantities are placed. If, for example, 3, 8 and 5 are to be added together, they are written 3+8+5; which is read 3 plus 8 plus 5.

If one quantity is to be subtracted from another, the operation is indicated by placing the sign of subtraction, *minus*, (—) after the minuend and before the subtrahend, thus: 8—5, which is read 8 minus 5.

To show that two or more factors are to be multiplied together, they are written in any order, and between them the sign of multiplication ($\times$), or sometimes a dot (·) is used, thus 8$\times$5, or 8 · 5 means 8 times 5, or 8 multiplied by 5.

If we have 4 multiplied by 5 and this product by 3 and the resulting product again by 2, we write: 4$\times$5$\times$3$\times$2, or 5$\times$2$\times$3$\times$4, or 2$\times$3$\times$4$\times$5, it being quite immaterial in this case as to the order in which the factors are placed.

To indicate that one quantity is to be divided by another, a colon (:) is placed between the dividend and the divisor (the divisor always being placed last), or instead of the colon, the sign(÷) is sometimes used. There is also another method of indicating division, and this must be remembered as most important; the dividend has a horizontal line drawn under it and the divisor is written under the line. Thus we have either 10 : 5, or 10÷5, or we may write $\frac{10}{5}$, the expression in all three cases being read 10 divided by 5.

The latter form will be met again in discussing the subjec of fractions, but it must not be forgotten that whatever other name may be given, it always means that the quantity above the line is to be divided by the quantity beneath.

15.

A quantity composed of several parts united by the signs + or —, may be called a compound expression in general, or named by the number of such parts or members which it contains ; numbers connected by the signs of multiplication or division being considered as single numbers. For example :

$$4\times8\ ;\ \frac{12}{3}\ ;\ \frac{4\times6}{2}\ \text{are single members.}$$

$$7+2\ ;\ \frac{64}{8} - 2\times3\ ;\ \text{expressions composed of two members}$$

each ; and

$$\frac{16}{2} + \frac{18}{3} - \frac{4\times9}{3}\ \text{is an expression composed of three members.}$$

16.

In order to indicate that any two expressions are equal to each other in value, the sign of equality ($=$) is placed between them, and such an indication of equality between two expressions is called an *equation*. The expression which is written before the sign of equality is called the *first* or *left-hand* side of the equation, while that which follows the sign of equality is called the *second*, or *right-hand* side, and any quantities on either side which are connected by the signs $+$, or $-$, are called *members*. For example :

$$8+5=13\ ;\qquad 2\times3\times4\times5=120\ ;$$
$$8-5=3\ ;\qquad \frac{64}{8} - 2\times3=5-3\ ;$$
$$\frac{3\times8}{6}=4\ ;\qquad \frac{4\times6}{2}+8=4\times6-\frac{8}{2}\ .$$

17.

When two quantities are not equal to each other, the fact may be indicated by using the sign $<$, which is placed between the two quantities in such a manner that the point is towards the smaller quantity. For example: $9>7$, or $7<9$ is read, 9 is greater than 8, or 7 is smaller than 9. In like manner, $3+6>8$, is read 3 plus 6 is greater than 8.

18.

If equal magnitudes be added to both sides of an equation, or if both sides be multiplied or divided by the same quantity, the equation will evidently still remain true; that is, the two sides will still remain equal to each other. Hence we have the rule regarding equations that : *Equal quantities treated in the same manner still remain equal to each other.*

Self-evident as this statement appears, it is most important, and should always be kept in mind when equations of any sort are under consideration.

For example : if we have the equation

$$2 \times 6 = 8 + 4$$

and add 5 to each side we have

$$2 \times 6 + 5 = 8 + 4 + 5$$

and the truth of the equation is still evident.

19.

When it is desired to indicate that a quantity composed of several members is to be taken several times ; i. e. is to be multiplied by another quantity, the quantity which is composed of several members is enclosed between parentheses, and the multiplier is placed before or after the parenthesis. In this case there is no need for the use of a multiplication sign and it is therefore omitted, the presence of the parenthesis marks sufficiently indicating that multiplication is intended. It makes no difference whether the members within the parenthesis are first added together and then multiplied by the quantity without, or whether each member is multiplied by it separately and the products afterwards added together, as the result will be the same in both instances. In order to see the truth of this statement it is only necessary to consider the quantity within the parenthesis as added to itself repeatedly as many times as there are units in the multiplier. For example : Suppose we have the quantity $3+5+1$ to be taken five times ; this is indicated by writing $5(3+5+1)$; which means :

$$= 15 + 25 + 5 = 45$$

which is the same as

$$5(3+5+1) = 5 \times 3 + 5 \times 5 + 5 \times 1$$

$$\text{or: } 5(3+5+1) = \begin{cases} 3+5+1 \\ 3+5+1 \\ 3+5+1 \\ 3+5+1 \\ 3+5+1 \end{cases}$$

$$5(3+5+1) = 5 \times 3 + 5 \times 5 + 5 \times 1$$
$$= 15 + 25 + 5 = 45$$

20.

When a magnitude composed of several members is to be divided by another magnitude, we see from the preceding section, that it is all the same whether each member is divided singly, or whether all the members are first collected together and the division performed afterward. For example : if we have the quantity 15+25+5 to be divided into 5 parts, which is indicated by writing $\frac{15+25+5}{5}$, (see § 14,) it is equal to :

$$\frac{15+25+5}{5} = \frac{15}{5} + \frac{25}{5} + \frac{5}{5} = 3+5+1 = 9$$

21.

When several numbers can each be divided by the same number without a remainder, then their sum is also divisible by the same number without a remainder. For instance, we have 15, 25, and 5, each divisible by 5 without remainder, hence their sum, 45, is also divisible by 5 without remainder.

22.

A product composed of several factors is evidently divisible without remainder by any one of the factors of which it is composed, and the quotient will be equal to the product of the other factors. For example : $2 \times 3 \times 4 \times 5 = 120$, and 120 divided by any one of these four factors will give a quotient which is equal to the product of the other three factors. We see also that 120 is exactly divisible by the products of any two or three of the factors, as $2 \times 4 = 8$, $3 \times 5 = 15$. We may thus see that

$$\frac{2 \times 3 \times 4 \times 5}{2 \times 3 \times 4} = 5 ; \qquad \frac{3 \times 2 \times 5 \times 4}{3 \times 5} = 2 \times 4$$

23.

When two numbers, such as 18 and 52 are to be *multiplied* together, and the product divided by a third number, such as 6, the operation is indicated thus : $\frac{18 \times 52}{6}$, which means that *only one* of the factors is to be divided, and the quotient multiplied by the other factor.

In the above example this would be equal to

$$\frac{18 \times 52}{6} = \frac{18}{6} \times 52 = 3 \times 52 = 156$$

In order to see the reason for this rule we have only to consider the number 52 added to itself 18 times, and the result divided by 6, and we see that this is the same as 52 added to itself 3 times.

Examples:

$$\frac{24 \times 36}{12} = 2 \times 36 = 72$$

$$\frac{18 \times 49}{7} = 18 \times 7 = 126$$

In such examples it is often desirable to be able to determine readily if one number is divisible by another without remainder. For this purpose the following easily remembered facts will be found useful.

24.

A number is always divisible by 2 when its last figure can be divided by 2, as 10; 24; 210; 506, etc., divisible by 2 × 2 or 4, when its last *two* figures are divisible by 2, as 100; 316; 5124; 500, etc.; by 2 × 2 × 2, or 8 when its last *three* figures are divisible by 2, as 5832; 1008; 2160, etc.; by 2 × 2 × 2 × 2, or 16, when its last *four* figures are divisible by 2, and so on indefinitely.

A number is divisible by 5 when its last figure is so divisible, as 10; 65; 75; 310, etc.; by 5 × 5 or 25 when its last *two* figures are so divisible, by 5 × 5 × 5 or 125, when its last *three* figures are so divisible, etc., etc.

A number is divisible by 3 or 9, when the sum of its figures is so divisible; for example, 141 is divisible by 3 because 1 + 4 + 1 = 6 is divisible by 3, and in like manner, 99; 111, 1101, 6594, etc. are so divisible. The number 5121 is divisible by 9, because the sum of its several figures 5 + 1 + 2 + 1 = 9, and for the same reason the numbers 99; 7074; 9297; 7992, etc., are divisible by 9.

Similar rules may be found for the easy factoring by other numbers such as 7, 13, 17, etc., but they are not so simple and are not of practical use. Tables of factors have also been made, giving directly the factors of numbers, and are used by those who are engaged in mathematical computations.

25.

All numbers which are divisible by 2 without a remainder, such as 2, 4, 184, 100, etc., are called *even* numbers, and all numbers which when divided by 2 leave a remainder of 1, are called *odd* numbers. Although 0 is really used as a symbol of position, and 1 a symbol of unity, yet it has become the custom to consider them also as numbers, 0 being even and 1 being odd.

A number which is divisible by any other number without a remainder, is capable of being divided into factors, and is called a *Composite* number, and the numbers into which is may be divided are called its factors. For example: 8 is a composite number, and its factors are 2 and 4; also 9, 10, 12, 27, etc., are composite numbers, since $9 = 3 \times 3$; $10 = 2 \times 5$; $12 = 3 \times 4 = 3 \times 2 \times 2$.

26

A number which is *not* divisible by any other number (excepting 1, and itself), cannot be separated into factors, and is called a *Prime* number. Such are 2, 3, 5, 7, 11, 13, 17, 19, 23, 29, 31, 37, 41, 43, 47, 53, etc.

With the exception of 2, all prime numbers are odd, but it by no means follows that all odd numbers are prime numbers.

In spite of the labors of mathematicians and the discovery of many curious properties of numbers, no general rule for the detection of prime numbers has been found. Extensive tables of prime numbers have been computed, and to them reference is made by those who have to make many computations involving factors.

27.

When several numbers are all exactly divisible by any one number, the latter number is called *their common divisor*, and the numbers which possess a common divisor are said to be *commensurable*. Thus for example: 21, 28, 14, 7 are commensurable with regard to each other, as are also 9, 27, 18; the first group have 7 as a common divisor, the latter group may be divided by 3 and by 9.

Numbers which have no common divisor, so that there is no number by which they can all be exactly divided are said to be prime to each other. Such, for example, are 8 and 9, 17 and 31.

28.

In order to find all the factors of a given number, both composite and prime, by which it can be divided without a remainder, we divide it first by a prime number, then the quotient by a prime number, and so on until the quotient becomes unity, and the divisors will be the *prime factors*. By multiplying these together (see § 22), we obtain the composite factors. The method and most convenient arrangement will be seen by taking an example. Let it be required to find the factors of 210.

The number is first written and a vertical line drawn to its right, and to the right of the line is placed the first prime divisor, and to the left, the quotient obtained by the divisior ; this quotient is again divided, and the operation continued until no more prime factors are found, thus :

EXAMPLE 1.

$$
\begin{array}{r|l}
210 & \\
105 & 2, \\
35 & 3, 6; \\
7 & 5, 10, 15, 30, \\
1 & 7, 14, 21, 42, 35, 70, 105
\end{array}
$$

Hence : $210 = 2 \times 105$
$210 = 2 \times 3 \times 35$
$210 = 2 \times 3 \times 5 \times 7$

Here we have first divided by the successive prime numbers which on trial were found to leave no remainder; we then multiplied 2 by 3, and obtained 6 ; then 2 by 5, then 2 by 7 ; then multiplied in like manner each of the prime factors by all the other prime factors, and so obtained this complete list of composite factors.

EXAMPLE 2. find all the factors, both prime and composite, of the number 360.

$$
\begin{array}{r|l}
360 & \\
180 & 2 \\
90 & 2, 4 \\
45 & 2, 4, 8 \\
15 & 3, 6, 12, 24 \\
5 & 3, 6, 12, 24, 9, 18, 36, 72 \\
1 & 5, 10, 20, 40, 15, 30, 60, 120, 45, 90, 180.
\end{array}
$$

We see by the above examples that 210 is capable of being factored into the prime numbers 2, 3, 5, 7, also that according to § 22, it is capable of being divided by 2, 3, 5, 7, 6, 10, 15, 30, 14, 21, etc., without remainder; all the numbers being products of two or more of the prime factors.

In the same manner we find that 360 is composed of the prime numbers 2, 2, 2, 3, 3, 5, and is divisible by 2, 3, 5, 4, 8, 6, 12, etc., etc., without remainder.

EXAMPLES. Find the prime and composite factors of the following numbers:

(1) 4158.

(2) 1836.

(3) 1155.

Answers. (1) Prime factors: 3, 3, 3, 2, 7, 11. Composite factors: 9, 27, 6, 18, 54, 21, 63, 189, 14, 42, 126, 378, 33, 99, 297, 22, 66, 198, 594, 71, 231, 693, 2079, 154, 362, 1386.

(2) Prime factors: 2, 2, 3, 3, 3, 17. Composite factors: 4, 6, 12, 9, 18, 36, 27, 54, 108, 34, 68, 51, 102, 204, 153, 306, 612, 459, 918.

(3) Prime factors: 5, 3, 7, 11. Composite factors: 15, 35, 55, 21, 33, 77, etc.

29

Those factors which are common to any two numbers, as, for example, 68, and 88, are very readily found if we first separate each of them into their prime factors, according to the preceding section. We thus find that $68 = 2 \times 2 \times 17$; and $88 = 2 \times 2 \times 2 \times 11$, from which it is evident that the numbers 68 and 88 are both divisible exactly by 2 and by $2 \times 2 = 4$. If, however, we wish to find the *greatest* common factor which will exactly divide two given numbers, it will be shorter to proceed in the following manner.

Divide the greater of the two numbers by the lesser, then divide the lesser by the remainder, and so keep on dividing each divisor by the remainder. As soon as a divisor is found which gives *no* remainder, it will be the greatest factor which is common to both of the original numbers. For example:

Example (1). 68)88(1
              68
              ――
           20)68(3
              60
              ――
              8)20(2
               16
               ――
               4)8(2
                 8

Example (2). 306)595(1
              306
              ―――
           289)306(1
              289
              ―――
              17)289(17
                 17
                 ――
                119
                119

In example (1) the greatest factor which will divide 68 and 88 is shown to be 4 ; while in example (2) the greatest common to 595 and 306 is 17 ; these being the first divisions which leave no remainder. (See Appendix, § 315.)

Example (3). Find the greatest common factor of 235 and 564.

Example (4). Find the greatest common factor of 1240 and 372.

Example (5). Find the greatest common factor of 65 and 112.

Answers. (3) 47.

(4) 124.

(5) The numbers have no common factor, and are therefore prime to each other.

30.

In the study of addition and subtraction of fractions, in subsequent pages, we shall sometimes find it necessary to determine the smallest number which is divisible by several other numbers without a remainder ; or as it usually termed : the *least common multiple*, of several given numbers.

This can readily be done in the following manner. First set aside those of the given numbers which are already contained in others ; if then among the remaining numbers two or more are found which possess a common factor, these can be divided by this common factor and instead of each of these numbers place the factor and the corresponding quotients. If then these are all multiplied together, and by the numbers which have no common factor, we get the least common multiple of the given numbers.

Suppose, for example, it is required to find a number which shall be divisible by 2, 5, 4, 18, 6, 9, 10, 24, 35 and 21, it is evident that we could find such a number merely by multiplying them all together. This, however, would give a very large number, and if we desire to find the least number in which they would all be contained, we must proceed otherwise.

First, we see at once by examining the numbers, that 2, 4, 5, 6 and 9 are contained in 18, 10, 24 and 35, and hence any number divisible by these latter will undoubtedly be divisible by the former.

The remaining numbers are then written in a row, (a) and it is seen that of them, 18 and 24 are divisible by 6 ; performing this division, we obtain a second row (b), here we see that the 3 is contained in 21, so it can be stricken out, and also see that two of the numbers are divisible by 5. This gives us the third row (c), in which the 2 and 7 are stricken out, being contained in 4 and in 21 ; and finally multiplying together the figures which remain, and the factors together, we get 2,520 for the least common multiple of the given numbers. The work in detail is as follows :

		2,	5,	4,	18,	6,	9,	10,	24,	35,	21
.	(a)				6)	18,	10,		24,	35,	21,
	(b)				5)	3,	10,		4,	35,	21,
	(c)					2,			4,	7,	21,

$$6 \times 5 \times 4 \times 21 = 2,520.$$

In the same manner we find 5,040 as the least common multiple of the numbers, 6, 9, 5, 7, 21, 56, 8, 12, 10 and 16.

	6,	9,	5,	7,	21,	56,	8,	12,	10,	16.
3)		9,		21,	56,		12,	10,	16,
8)		3,		7,	56,		4,	10.	16,
		3,		7,				10,	2,

$$3 \times 8 \times 3 \times 7 \times 10 = 5,040.$$

EXAMPLES :

What is the least common multiple of 2, 11, 9, 21, 8, 18, 7, 22 ?   Answer, 5,544.

What is the least common multiple of 2, 3, 4, 6 ? Ans. 12.

What is the least common multiple of 5, 12, 20, 15 ? Ans. 60.

What is the least common multiple of 124 and 100 ? Ans. 3,100.

BOOK IV.

Common or Vulgar Fractions.

31.

We have already seen that in order to express a value for any magnitude it must be compared with some unit of the same kind. At the same time it often occurs that the unit is not contained in the given magnitude an exact number of times, and indeed the quantity to be measured may itself be smaller than the unit. In such cases the way out of the difficulty is to conceive the unit as divided up into smaller units, and use the smaller unit as a measure. If we have a clear idea of the value of the original unit, we must also have just as clear a conception of any definite subdivisions of it.

When, therefore, we find that the original unit is not contained exactly in the quantity to be measured, we may sub-divide the unit into any definite number of parts which will best enable us to express the value of the excess.

For this purpose we may take the 3rd, 8th, 100th or any subdivision of the original unit, as may be found most convenient.

We may thus express the value of the given quantity by stating the number of times the original unit is contained in it, and also the number of times the sub-division is contained in the excess, and so give a clear and definite value to any magnitude.

The original unit we may consider as the *whole* unit and the sub-division, (or smaller unit) as a portion, or *fraction* of the whole unit; hence we call that portion of the quantity which is measured by the whole unit, a *whole number*, and that portion which is measured by the fraction of the unit, the *fraction ;* both taken together being called a *mixed number*.

(The fraction really may be considered a whole number, if by the word "fraction" we understand simply a name for a smaller unit.)

32.

In order to be able to express these quantities clearly and briefly by means of figures, the following notation has been adopted. The whole unit and the whole number are written as usual, and the fractional portion immediately following ; the number of fractional units being written above a short horizontal line, and the number of such units contained in a single whole unit written below the line. The fractional unit is written alone by placing a 1 above the line and the number of fractional units in a whole unit below the line.

Thus if the fractional unit is the eighth part of a yard it is written $\frac{1}{8}$, and if the quantity to be measured is as long as five of these eighths, it is written $\frac{5}{8}$ yd. The number which is placed above the line, and which gives the number of fractional units, is called the *numerator* ; the number below the line, which gives the denomination of the fraction, or tells of what portion of the whole unit it consists, is called the *denominator*. For example, in the fraction $\frac{5}{8}$ (read five-eighths), 5 is the numerator, 8 the denominator, and $\frac{1}{8}$ is the fractional unit. The 8 tells that it takes eight of the fractional units to equal one whole unit, and the 5 tells that 5 of these fractional units (or eighths) are equal to the value to be expressed. In like manner, in $\frac{15}{34}$, the fractional unit is $\frac{1}{34}$, 15 is the numerator, and 34 the denominator.

A *mixed number* is expressed by writing the fraction immediately after the whole number, thus $7\frac{5}{8}$ (read seven and five-eighths) is equal to seven whole units, and five times the one-eighth part of a whole unit, and is the same as $7+\frac{5}{8}$. See Appendix, § 317.

33.

The form of a fraction coincides with one of the expressions to indicate division, as already explained in § 14. This is quite correct, and every fraction should also be considered as a statement of division. It is evidently just the same thing to say that $\frac{3}{4}$ pound is three times the fourth part of a pound, or one time the fourth of three pounds. Or, taking another case, 1 kilogram= equals 1000 gram ; and 3 kilos=3×1000 grams=3000 grams. Now it is evidently exactly the same value to take the fourth part of 1000 grams, 3 times, as to take the fourth part of 3000 grams 1 time, for $3 \times \dfrac{1000}{4} = \dfrac{3 \times 1000}{4} = 750$ (see § 23).

It is therefore as correct to say $\frac{8}{4}$ means 8 divided by 4 as to say eight-fourths. Any fraction in which the numerator and denominator are alike, as $\frac{4}{4}$, $\frac{3}{3}$, $\frac{7}{7}$, $\frac{100}{100}$ is equal to unity, and eight-fourths is equal to twice $\frac{4}{4}$ or equal to two whole units.

This conception of a fraction as an expression of the division of the numerator by the denominator is most important, and and should always be kept in mind in view of what follows.

34.

Any whole number may be written as a fraction of any desired denomination simply by multiplying the number by the desired denominator and placing the product over this denominator. If, for example, we wish to express 7 in the form of a fraction with a denominator of 4 or 5, we may write:

$$7 = \frac{7 \times 4}{4} = \frac{28}{4}; \text{ or } 7 = \frac{7 \times 5}{5} = \frac{35}{5}$$

In the same manner any mixed number can be converted into a fraction by multiplying the whole number by the denominator of the fractional portion, adding the numerator of the fraction to the product and placing the sum over the denominator.

For example : $7\frac{5}{8} = \dfrac{7 \times 8 + 5}{8} = \dfrac{61}{8};$ $5\frac{3}{4} = \dfrac{23}{4};$ $2\frac{2}{3} = \dfrac{8}{3};$ $200\frac{11}{13} = \dfrac{2611}{13}$, etc.

All fractions of which the numerator is greater than the denominator, as $\frac{41}{8}$, $\frac{8}{3}$, etc., are called *improper* fractions, while those in which the numerator is smaller than the denominator, as $\frac{1}{3}$, $\frac{5}{8}$, etc., are called *proper* fractions.

35.

The two cases in which fractions are necessary for the measurement and expression of magnitudes are the following :

1. When the magnitude which is to be measured is a denominate quantity, and the unit will not divide it exactly. This

occurs so frequently, that the units in customary use are subdivided into smaller units for the purpose of enabling the expression of fractional parts to be readily made. Thus the pound is
subdivided into ounces, and these again into grains ; the kilogram
into 10ths, 100ths and 1000ths; each of these being given
names ; also measures of length are divided into sub-units, as
feet and inches, meters and millimeters, etc., etc. The smaller
units can again be converted into fractional parts of the larger
units, it only being necessary to know how many of the sub-
units are contained in the main unit.

2. For the comparison, measurement and division of abstract numbers, the fractional form of expression is much used.
In division we always endeavor to ascertain *how often* the divisior
is contained in the dividend, and hence we can always consider
the divisor as composed of a group of several equal units. We
then have possible the three following cases : First ; *how many
times* larger is one magnitude than another of a similar kind? In
this case the quotient is always an undenominate, abstract number, representing only the ratio or relation between the two compared magnitudes, and telling nothing whatever about their denomination or actual magnitude. For example, we may ask :
how many times greater is 20 than 4 ? or \$20 than \$4 ? or 20
pounds than 4 pounds ; and the answer in each case is 5, not
\$5, or 5 pounds, but 5 *times*, the result being an abstract number.
If the divisor is not contained in the dividend an exact number of
times, the remainder must be appended to the quotient in the
form of a fraction. Thus, if it be asked how many times greater
is 23 than 4 ? the answer is that 4 is contained in 23, 5 whole
times, and that the fourth part of the unit 1 is contained in the
remainder 3 times, hence $\frac{23}{4}=5\frac{3}{4}$, and 23 is $5\frac{3}{4}$ times greater than
4. In the same manner we have $\frac{99}{32}=3\frac{3}{32}$. If the question is
how many times is 8 contained in 3 ? the answer is $\frac{3}{8}$ times, and
in all these cases it will be seen how the fractional form is merely
the expression of division.

Second : It is required to reduce lower units to higher
ones. In this case the unit of the undenominate quotient takes
the appellation of the higher unit to which the divisor is equal.
For example : if it be asked how many pounds are equal to 160
ounces? Knowing that there are 16 ounces in one pound, we

have $\frac{160}{16}$ oz.$=$10 pounds ; or if the question be, how many kilograms are 556 grams, we have $\frac{556}{1000}$ kilograms, since 1 kilogram$=$1000 grams.

Third : It is required to divide a magnitude into a given number of parts. In this case the quotient is of the same denomination as the dividend. For example : the 4th part of $23, is 5\frac{3}{4}$; that is 5 dollars and 3 quarter dollars. The half of £5, is £2$\frac{1}{2}$=£2 10s.

If the student has fully understood the foregoing, and masters thoroughly the following five rules, he will be able to work with fractions quite as easily as with whole numbers.

36.

When the NUMERATOR *of a fraction is multiplied or divided by a number, the value of the fraction will become as many times greater or smaller, as there are units in the multiplier or divisor,*

For example : If the numerator of the fraction $\frac{6}{32}$ be multiplied by 3, the fraction becomes $\frac{18}{32}$. The first fraction contained the unit $\frac{1}{32}$ six times, and the latter contains the same unit eighteen times, hence it is evident that 18 units are 3 times as many as 6 units of the same kind. If we divide the numerator of the fraction $\frac{6}{32}$ by 3, we get $\frac{2}{32}$, and this is clearly only $\frac{1}{3}$ as much. In the same manner we see that $\frac{24}{63}$ is 6 times as much as $\frac{4}{63}$, and that $\frac{3}{63}$ is the eighth part of $\frac{24}{63}$, etc.

37.

When the DENOMINATOR *of a fraction is multiplied or divided by a number, the value of the fraction becomes* INVERSELY *smaller or greater, as many times as there are units in the multiplier or divisor.*

This rule will be seen by comparison to be just the reverse of the preceeding, and although perhaps not so readily grasped by the beginner, will nevertheless be quite clear on inspection.

It will be most readily understood by thinking of the fractional unit as a length. Suppose we have the fraction $\frac{3}{16}$ of an inch, and multiply the denominator by 4, we get $\frac{3}{64}$ of an inch. If now we only think of an inch as divided into 16ths and into 64ths, or, indeed, examine these divisions on a good drawing scale or foot-rule, we shall see that the fractional unit $\frac{1}{16}$ is four times as great as the fractional unit $\frac{1}{64}$ and hence that 3 times this smaller unit, $\frac{3}{64}$, is only $\frac{1}{4}$ as large as $\frac{3}{16}$. Hence our *multiplying* the denominator by 4 has made the value only $\frac{1}{4}$ as much as it was before the multiplication.

If we have a number of weights of 1lb. 2lbs. 3lbs. etc., we can see that a given number of 1lb. weights will only weigh $\frac{1}{2}$ as much as the same number of 2lb. weights, or $\frac{1}{3}$ as much as 3lb. weights, and that smaller the unit the less the value of a given number of units. The reverse is also now evident. If we divide the denominator of $\frac{3}{16}$ by 4 we get $\frac{3}{4}$, and this fraction is four times as great as $\frac{3}{16}$, because the unit $\frac{1}{4}$, is four times as great as the unit $\frac{1}{16}$. If we divide the denominator of the fraction $\frac{3}{100}$ of a dollar by 25 we get $\frac{3}{4}$ of a dollar, which is evidently 25 times as much ; the first is only 3 cents. The second is 75 cents.

38.

When the numerator and denominator of a fraction are BOTH *multiplied or divided by the same number, the value of the fraction remains unchanged.*

If, for example, we multiply both numerator and denominator of the fraction $\frac{4}{9}$ by 3 we obtain the form $\frac{12}{27}$, the unit $\frac{1}{27}$ being 3 times smaller than $\frac{1}{9}$, but since there are 3 times as many of them the value is the same, and $\frac{4}{9}=\frac{12}{27}$. In the same manner, if we divide both numerator and denominator of $\frac{4}{9}$ by 3 we get $\frac{3}{4}$, in which the unit $\frac{1}{3}$ is 3 times greater than $\frac{1}{9}$, but there are only one third as many, and $\frac{4}{9}=\frac{3}{4}$. We can then change any fraction into an indefinite number of forms without altering its value.

39.

By dividing the numerator and denominator of a fraction by their greatest common divisor (§ 29), or by repeatedly dividing them by any factors which they may have in common until they become prime to each other we may reduce the fraction to its lowest terms, and thus simplify the expression without changing its value. Thus the fraction $\frac{24}{84}$ divided above and below by 12 may be reduced to $\frac{2}{7}$, or by dividing by $2\times2\times3$, $\frac{24}{84}=\frac{12}{42}=\frac{6}{21}=\frac{2}{7}$.

Examples: Reduce the following fractions to their lowest terms : $\frac{7}{14}$, $\frac{4}{6}$, $\frac{34}{68}$, $\frac{111}{148}$, $\frac{104}{160}$, $\frac{111}{111}$, $\frac{288}{381}$, $\frac{4800}{7804}$, $\frac{72}{81}$, $\frac{6}{9}$, $\frac{8600}{9600}$, $\frac{7141}{7161}$, $\frac{66}{88}$, $\frac{24}{84}$, $\frac{6}{8}$.

Answers: $\frac{1}{2}$, $\frac{2}{3}$, $\frac{1}{2}$, $\frac{37}{313}$, $\frac{2}{47}$, $\frac{837}{3037}$, $\frac{288}{381}$, $\frac{100}{126}$, $\frac{8}{9}$, $\frac{2}{3}$, $\frac{127}{127}$, $\frac{27}{21}$, $\frac{66}{88}$, $\frac{7}{18}$, $\frac{6}{8}$.

40.

If it is desired to reduce a fraction to a form having a given denominator, without altering its value, it is only necessary to make the new numerator as many times greater than the old numerator as the new denominator is greater than the old denominator (see § 38). Hence it is only necessary to divide the new denominator by the old one and multiply the quotient by the old numerator, and the product will be the new numerator. For instance, if the fraction $\frac{4}{5}$ is to be reduced to one having a denominator of 45, we have $\frac{4}{5}=\frac{36}{45}$. In the same manner $\frac{2}{3}$ reduced to a denominator 12, gives $\frac{8}{12}$, etc.

Examples. Bring the following fractions to the indicated forms : $\dfrac{5}{6}=\dfrac{\cdots}{144}$; $\dfrac{6}{17}=\dfrac{\cdots}{51}$; $\dfrac{5}{7}=\dfrac{\cdots}{84}$; $\dfrac{9}{13}=\dfrac{\cdots}{2613}$

Answers : $\dfrac{5}{6}=\dfrac{120}{144}$; $\dfrac{6}{17}=\dfrac{18}{51}$; $\dfrac{5}{7}=\dfrac{60}{84}$; $\dfrac{9}{13}=\dfrac{1809}{2613}$

41.

In order that several fractions with various denominators may be added together, they must first all be reduced to the same denominator or fractional unit. This is not difficult to accomplish, since we may, according to § 30, easily find a number which will contain all of the denominators without a remainder. Suppose we have, for example, to reduce all the following fractions to the same denominator (called reduction to a common denominator), $\frac{2}{3}$, $\frac{1}{4}$, $\frac{5}{6}$, $\frac{7}{9}$, $\frac{11}{12}$, $\frac{5}{8}$.

We find by § 30, that the least common multiple of 3, 4, 6, 9, 12 and 8 is 72, and therefore have : $\frac{2}{3}=\frac{48}{72}$; $\frac{1}{4}=\frac{18}{72}$; $\frac{5}{6}=\frac{60}{72}$; $\frac{7}{9}=\frac{56}{72}$; $\frac{11}{12}=\frac{66}{72}$; $\frac{5}{8}=\frac{45}{72}$.

Addition.

42.

We have already seen that only such numbers can be added together as are based upon the same unit. If, therefore, it is required to add fractions which have not a common denominator, they must first be reduced to a common denominator. This being done, it is only necessary to add the numerators together and set the sum over the common denominator. If mixed numbers are to be added together, the whole numbers and the fractions are added separately.

For example:

$$\tfrac{3}{8}+\tfrac{2}{8}=\tfrac{5}{8}\ ;\qquad \tfrac{4}{17}+\tfrac{5}{17}+\tfrac{9}{17}+\tfrac{1}{17}=\tfrac{18}{17}=1\tfrac{1}{17}\ ;\qquad 3\tfrac{3}{17}+2\tfrac{6}{17}=\tfrac{18}{17}.$$

The method of reduction to a common denominator, applied to a number of fractions which are to be added together, is shown in the following example, which gives a convenient arrangement.

Let it be required to find the sum of:

$$\tfrac{2}{3}+\tfrac{3}{4}+\tfrac{5}{6}+\tfrac{7}{8}+\tfrac{14}{27}+\tfrac{4}{9}+\tfrac{7}{15}+\tfrac{5}{108}+\tfrac{7}{54}$$

First finding the least common multiple of the denominators, we have according to § 30 :

$$
\begin{array}{c|ccccccccc}
 & 3, & \not{4}, & \not{6}, & 8, & \not{27}, & \not{9}, & 15, & 108, & \not{54} \\
\hline
3) & 3, & 8, & 15, & & 108 \\
\hline
4) & 1, & 8, & 5, & & 36 \\
\hline
 & 2, & 5, & & 9
\end{array}
$$

hence the least common multiple is: $3\times4\times2\times5\times9=1080$, and the corresponding numerators are :

$$1080=3\times4\times2\times5\times9.$$

$$
\begin{array}{c|c}
\tfrac{2}{3} & 720 \\
\tfrac{3}{4} & 810 \\
\tfrac{5}{6} & 900 \\
\tfrac{7}{8} & 945 \\
\tfrac{14}{27} & 560 \\
\tfrac{4}{9} & 480 \\
\tfrac{7}{15} & 504 \\
\tfrac{5}{108} & 50 \\
\tfrac{7}{54} & 140
\end{array}
$$

$$\tfrac{5109}{1080}=4\tfrac{789}{1080}=4\tfrac{263}{360}$$

The numerators are found by dividing the new denominator, 1080, by the old denominators and multiplying by the old numerators. Thus for the first fraction $\tfrac{2}{3}$ we have $1080\times\tfrac{2}{3}=360\times2=720$, etc. The results may often be obtained more readily if the factors of the common denominator are used as :

$$\frac{1080}{15}=\frac{2\times3\times4\times5\times9}{3\times5}=2\times4\times9=72 \text{ and } 72\times7=504=\text{the new}$$

numerator for $\tfrac{7}{15}=\tfrac{504}{1080}$.

For another example let it be required to find the sum of :

$$\tfrac{3}{4}+\tfrac{5}{7}+\tfrac{11}{12}=\frac{63+60+77}{7\times12}=\frac{200}{84}=2\tfrac{8}{21}.$$

Here we see that the common denominator will be $7\times12=84$, since 4 is a factor of any multiple of 12, and the numerators are obtained as before.

Mental calculations, which should be used wherever possible, are often available in operations with fractions. For instance, the sum of $\tfrac{3}{4}+\tfrac{8}{24}=\tfrac{26}{24}=1\tfrac{1}{12}$, is readily obtained mentally, since we see at once that we have only to multiply both numerator and denominator of the first fraction by 6 to bring it to the same denominator as the other one, after which the addition is simply performed. Again, the sum $\tfrac{5}{6}+\tfrac{7}{8}=\tfrac{41}{24}=1\tfrac{17}{24}$, is obtained by multiplying the first fraction by 4 and the second by 3, to reduce them to 24ths.

Again : $\tfrac{3}{7}+\tfrac{1}{4}=\tfrac{19}{28}$; obtained by multiplying by 4 and 7 to obtain 28ths.

Again : $\tfrac{2}{9}+\tfrac{7}{8}=\dfrac{16+63}{8\times9}=\tfrac{79}{72}=1\tfrac{7}{72}$;

Again : $\tfrac{1}{2}+\tfrac{3}{4}+\tfrac{5}{8}=\tfrac{15}{8}=1\tfrac{7}{8}$; the first and second factors have been reduced by multiplying them respectively by 4 and 2.

Sometimes the work can be simplified by performing it in two operations as : $\tfrac{1}{2}+\tfrac{3}{4}+\tfrac{7}{8}+\tfrac{2}{3}+\tfrac{7}{9}=2\tfrac{1}{8}+1\tfrac{4}{9}=3\tfrac{41}{72}.$

Here the first three and the last two fractions are added separately, and the sums then added together.

Examples : Find the sum of

(1) $\tfrac{3}{4}+\tfrac{5}{7}+\tfrac{11}{12}+\tfrac{7}{8}+\tfrac{5}{21}+\tfrac{1}{3}.$ Answer. $3\tfrac{33}{56}$

(2) $12\tfrac{2}{3}+3\tfrac{1}{2}+4\tfrac{7}{15}+\tfrac{1}{4}+2\tfrac{8}{15}+\tfrac{2}{3}.$ " $23\tfrac{23}{30}$

(3) $\tfrac{11}{18}+\tfrac{3}{8}+\tfrac{7}{4}+\tfrac{8}{19}+\tfrac{5}{4}+\tfrac{11}{12}+\tfrac{5}{8}+\tfrac{9}{12}.$ " 7

(4) $\tfrac{15}{124}+\tfrac{3}{63}+2\tfrac{1}{8}\tfrac{}{8}+\tfrac{5}{12}.$ " $\tfrac{227}{312}$

(5) $\tfrac{3}{7}+\tfrac{5}{9}+1\tfrac{1}{14}+\tfrac{5}{18}+3\tfrac{2}{3}+1\tfrac{4}{126}.$ " $5\tfrac{4}{126}$

(6) $\tfrac{5}{16}+\tfrac{3}{8}+1\tfrac{17}{24}+\tfrac{9}{11}+\tfrac{3}{8}+\tfrac{8}{8}+\tfrac{9}{11}.$ " $4\tfrac{39}{240}$

(7) $\tfrac{1}{10}+\tfrac{2}{8}+1\tfrac{3}{100}+\tfrac{1}{1000}+1\tfrac{3}{100}+\tfrac{1}{8}.$ " $\tfrac{761}{1000}$

Subtraction.

43.

The same principle applies in Subtraction as in Addition. If the fractions all have not the same denominator they must first be reduced to a common denominator before the subtraction can be performed. The numerator of the subtrahend is then subtracted from the numerator of the minuend, and the remainder is written over the common denominator. For example :

$$\tfrac{5}{8}-\tfrac{3}{8}=\tfrac{2}{8}=\tfrac{1}{4} \; ; \; \tfrac{3}{4}-\tfrac{2}{3}=\tfrac{9}{12}-\tfrac{8}{12}=\tfrac{1}{12}$$

If a fraction is to be subtracted from a whole number, a unit must be taken from the whole number and converted into a fraction of the same denominator as the subtrahend. For example :

$$5-\tfrac{3}{7}=4\tfrac{7}{7}-\tfrac{3}{7}=4\tfrac{4}{7}$$

Again : $16-\tfrac{12}{15}=15\tfrac{15}{15}-\tfrac{12}{15}=15\tfrac{3}{15}$

If mixed numbers are to be subtracted the fractional portions are first reduced to a common denominator and the fractions and whole numbers are subtracted separately. If the fraction in the subtrahend is larger than that in the minuend, a unit must be taken from the whole number of the minuend and converted into a fraction of the same denominator. For example :

$$5\tfrac{7}{8}-1\tfrac{3}{4}=5\tfrac{21}{24}-1\tfrac{18}{24}=4\tfrac{5}{24}$$

Again : $14\tfrac{2}{3}-2\tfrac{7}{8}=13\dfrac{24+16}{24}-2\tfrac{21}{24}=11\tfrac{19}{24}$

In this last example $\tfrac{2}{3}=\tfrac{16}{24}$ and $14=13\tfrac{24}{24}$, so that $14\tfrac{2}{3}=13\tfrac{40}{24}$.

It is more convenient to subtract the 16 from the 24 and add the remainder 3 to the 16, than to subtract the 21 from the 40, etc. Examples :

$$\tfrac{3}{4}-\tfrac{1}{2}=\tfrac{1}{4}\;; \quad \tfrac{3}{5}-\tfrac{3}{4}=\tfrac{7}{20}\;; \quad \tfrac{3}{4}-\tfrac{5}{8}=\tfrac{1}{8}\;; \quad \tfrac{5}{9}-\tfrac{1}{2}=\tfrac{1}{18}\;; \quad 3-\tfrac{3}{4}=2\tfrac{1}{4}\;;$$
$$5\tfrac{5}{12}-2\tfrac{11}{12}=2\tfrac{1}{2}\;; \quad 22\tfrac{5}{7}-4\tfrac{3}{4}=17\tfrac{27}{28}\;; \quad 2\tfrac{12}{41}-1\tfrac{3}{4}=1\tfrac{77}{123}\;; \quad \tfrac{5}{8}-\tfrac{5}{8}=\tfrac{5}{24}\;;$$
$$\tfrac{1}{4}-\tfrac{1}{8}=\tfrac{1}{6\cdot6}\;; \quad \tfrac{3}{8}-\tfrac{3}{9}=\tfrac{1}{12}\;; \quad \tfrac{1}{5}-\tfrac{1}{10}=\tfrac{1}{30}\;; \quad 4\tfrac{8}{15}-2\tfrac{8}{9}=1\tfrac{44}{45}.$$

The answers are given to these examples, but in these and all other examples the student must work out the operations and reductions in full.

Multiplication.

44.

The rule for multiplication of two fractions is easily remembered. *Multiply numerator by numerator and denominator by denominator.*

This rule, which will be explained below, holds good in all cases. If either one of the factors be a whole number it may be assumed to have a denominator of unity, and if either or both of the factors be mixed numbers they should be reduced to improper fractions before proceeding with the multiplication.

Examples :

$$(1)\quad \tfrac{1}{5}\times\tfrac{7}{9}=\frac{1\times7}{5\times9}=\tfrac{7}{45}$$

$$(2)\quad \tfrac{4}{5}\times\tfrac{7}{9}=\frac{4\times7}{5\times9}=\tfrac{28}{45}$$

$$(3)\quad 4\times\tfrac{7}{8}=\tfrac{4}{1}\times\tfrac{7}{8}=\tfrac{28}{8}=3\tfrac{1}{2}$$

$$(4)\quad 2\tfrac{3}{4}\times\tfrac{7}{8}=\tfrac{11}{4}\times\tfrac{7}{8}=\tfrac{77}{32}=2\tfrac{5}{36}$$

$$(5)\quad 2\tfrac{3}{4}\times2\tfrac{7}{9}=\tfrac{11}{4}\times\tfrac{25}{9}=\tfrac{275}{36}=7\tfrac{23}{36}$$

Explanation : Multiplication consists in taking one quantity (the multiplicand) as many times as unity is contained in the other quantity (the multiplier). If, for example, the multiplier is $\tfrac{1}{5}$ and the multiplicand is $\tfrac{7}{9}$, we see that the multiplier does not contain a whole unit, but only $\tfrac{1}{5}$ part of a unit, which it contains *once*. The multiplicand must therefore not be taken once, but only $\tfrac{1}{5}$ times. According to §37 we know that we will make $\tfrac{7}{9}$ equal to $\tfrac{1}{5}$ of its value by making the denominator five times larger, whence $\tfrac{1}{5}\times\tfrac{7}{9}=\tfrac{7}{45}$.

To the beginner it will appear that in this case $\tfrac{1}{5}$ times $\tfrac{7}{9}$, we have, in the form of multiplication, performed a *division*, since the value of the quantity $\tfrac{7}{9}$ is not increased but diminished. This apparent confusion of operations will easily be understood, and the difficulty cleared away if we remember that we can take a quantity a fractional part of a time, as well as we can take it a whole number of times. It is just the same thing if say we take the fifth part of a quantity, or say we take it $\tfrac{1}{5}$ times (multiply it by $\frac{1}{4}$), and thus we must think of multiplication in a broader sense than was attempted in § 9, so it may be defined thus :

Multiplication consists in taking a quantity as many times as another quantity contains units.

This general definition of multiplication involves corresponding general meanings for the terms multiplicand, multiplier and product, and considers, for instance, the word product as meaning a given number of times a fraction is taken, as well as a whole number. Multiplicand applies to any magnitude, whether either itself or any part of it is to be taken ; the multiplier shows how often the whole multiplicand, or any part of it, is to be taken. In this connection the rules in §§ 36 and 37 should be compared.

For example : If $\frac{4}{5}$ is the multiplier and $\frac{7}{8}$ the multiplicand, the multiplier contains the 5th part of a unit 4 times, and hence it must also contain $\frac{1}{5}$ part of the multiplicand 4 times.

According to § 37, the 5th part of $\frac{7}{8}$ is $\frac{7}{48}$. According to § 36 this is taken 4 times by multiplying the numerator 7 by 4, which gives $\frac{28}{40}$, whence $\frac{4}{5} \times \frac{7}{8} = \frac{28}{40}$. The operation of taking a quantity $\frac{4}{5}$ times, therefore, involves both multiplication and division.

45.

When it is required to multiply several fractions together, all the numerators are multiplied together to form a new numerator, and all the denominators multiplied together to form the new denominator. For example, if we have $\frac{8}{9}$ to be multiplied by $\frac{4}{7}$ and the product by $\frac{2}{5}$, and this resulting product again by $\frac{1}{3}$, we have :

$$\tfrac{8}{9} \times \tfrac{4}{7} \times \tfrac{2}{5} \times \tfrac{1}{3} = \frac{8 \times 4 \times 2 \times 1}{9 \times 7 \times 5 \times 3} = \tfrac{64}{945}$$

When the multiplier is a proper fraction the product will of course be less than the multiplicand. In like manner, if all the factors are proper fractions the product will always be smaller than any one of the factors. For instance, in the example $\frac{4}{5} \times \frac{4}{5} \times \frac{1}{3} = \frac{24}{105}$, we have $\frac{24}{105} < \frac{1}{3}$; $\frac{24}{105} < \frac{4}{5}$; or $\frac{24}{105} < \frac{4}{7}$.

46.

In multiplying several fractions together, we often find that the numerators of some of the fractions are the same as the denominators of others, or have common factors, and these can at once be divided out. This will not affect the result, as it is obviously the same thing whether a reduction be made before or after the multiplication is performed.

Beginners often neglect to make as much use of this labor-saving principle as they should

We see at once that $\frac{4}{7} \times \frac{7}{5} = \frac{4}{5}$, as the 7 in the numerator of one fraction is canceled by the 7 in the denominator of the other, and the result is read immediately. In the same manner $\frac{12}{17} \times \frac{17}{32}$ $= \frac{3}{8}$; $\frac{11}{13} \times \frac{5}{11} = \frac{5}{13}$; $6 \times \frac{2}{3} = 2 \times 2 = 4$; $\frac{3}{5} \times 25 = 3 \times 5 = 15$; $16 \times \frac{5}{32} = \frac{5}{2} = 2\frac{1}{2}$. If we have to multiply together the quantities $\frac{2}{3}$, $\frac{3}{4}$, $2\frac{3}{4}$, $\frac{5}{7}$, $\frac{4}{5}$ and $2\frac{1}{3}$ we have :

$$\frac{2 \times 3 \times 11 \times 5 \times 4 \times 7}{3 \times 4 \times 4 \times 7 \times 5 \times 3} = \frac{2 \times 11}{4 \times 3} = \frac{11}{6} = 1\frac{5}{6}$$

When one of two factors is mixed and the other a whole number, is more convenient to multiply the whole number and the fraction of the former by the latter separately and add the two products together afterwards. For example :

$$6 \times 5\frac{2}{21} = 30 + \frac{12}{21} = 30\frac{4}{7}$$
$$12 \times 13\frac{3}{4} = 156 + \frac{45}{8} = 165\frac{5}{8}$$

This method is also convenient when one factor is a mixed number and the other a proper fraction.

When both factors are mixed numbers, the whole number and fraction of each may be multiplied by the whole number and fraction of the other, and the four products then added together. This method will only be found preferable when there are not more than two fractions, or when the whole numbers are large. In most cases it will be found simpler to reduce them to improper fractions before multiplying.

Examples of both methods are here given :

$$25\frac{5}{8} \times 124\frac{3}{4} = 25 \times 124 + 25 \times \frac{3}{4} + \frac{5}{8} \times 124 + \frac{5}{8} \times \frac{3}{4}$$
$$= 3100 + 18\frac{3}{4} + 103\frac{1}{3} + \frac{5}{8}$$
$$= 3221 + \frac{3}{4} + \frac{5}{8} + \frac{1}{3} = 3222\frac{17}{24}$$

$$3\frac{3}{4} \times 6\frac{3}{4} = \frac{15}{4} \times \frac{34}{5} = \frac{3 \times 17}{2} = 25\frac{1}{2}$$

$$25\frac{5}{8} \times 124\frac{4}{5} = \frac{155}{6} \times \frac{624}{5} = 31 \times 104 = 3224$$

In connection with the multiplication of fractions occurs the problem of reduction of fractional parts of a higher denominate unit to a lower unit, since *this operation really consists in the multiplication of the fraction by the number of lower units contained in a higher unit.*

For instance, there are 100 cents in one dollar, hence $\frac{3}{4}$ of a dollar is equal to $\frac{3}{4} \times 100 = 3 \times 25 = 75$ cents.

The following examples should be carefully worked out with the aid of the principle of cancellation of common factors.

(1) $\frac{3}{8} \times \frac{5}{7} = \frac{3}{4}$

(2) $\frac{2}{6} \times \frac{7}{8} = \frac{7}{30}$

(3) $\frac{7}{8} \times \frac{4}{7} = \frac{1}{2}$

(4) $\frac{1}{2} \times \frac{3}{4} = \frac{3}{8}$

(5) $2 \times \frac{1}{2} = 1$

(6) $1 \times \frac{1}{2} = \frac{1}{2}$

(7) $\frac{1}{2} \times \frac{1}{2} = \frac{1}{4}$

(8) $\frac{1}{2} \times \frac{1}{3} = \frac{1}{6}$

(9) $\frac{3}{14} \times \frac{28}{41} = \frac{6}{41}$

(10) $\frac{22}{66} \times \frac{5}{11} = \frac{2}{13}$

(11) $\frac{5}{31} \times \frac{4}{8} = \frac{4}{31}$

(12) $\frac{11}{11} \times \frac{11}{66} = \frac{11}{13}$

(13) $125 \times 7\frac{4}{5} = 975$

(14) $\frac{3}{8} \times 11\frac{4}{11} = 6\frac{9}{11}$

(15) $\frac{6}{11} \times 2\frac{3}{4} = 1\frac{1}{4}$

(16) $2\frac{4}{6} \times 33\frac{1}{2} = 94\frac{11}{13}$

(17) $\frac{2}{3} \times 1\frac{1}{2} = 1$

(18) $\frac{3}{4} \times \frac{5}{6} \times \frac{4}{5} \times 6 = 3$

(19) $\frac{3}{8} \times \frac{4}{7} \times \frac{4}{5} = \frac{40}{180}$

(20) $2\frac{3}{4} \times \frac{4}{5} \times \frac{7}{9} \times 2\frac{3}{8} = 4\frac{76}{135}$

(21) $23\frac{5}{8} \times 22\frac{6}{13} = 533\frac{1}{2}$

(22) $33\frac{4}{8} \times 100 = 3366\frac{4}{8}$

(23) $27\frac{5}{8} \times \frac{7}{100} = 1\frac{7}{80}$

(24) $\frac{111}{233} \times \frac{28}{117} = \frac{11}{2131}$

Division.

47.

In indicating the division of fractions the sign $\div$ or its abbreviation : may be conveniently used.* For instance, to indicate that $\frac{4}{5}$ is divided by $\frac{2}{3}$, it is better to write $\frac{4}{5} : \frac{2}{3}$, than $\dfrac{\frac{4}{5}}{\frac{2}{3}}$

The division of fractions may be considered as a special case of multiplication, and converted into a multiplication.

The very simple rule for performing a division of fractions is as follows :

To divide one fraction by another fraction, invert the divisor (placing the numerator below the line, and the denominator above) and then multiply the dividend by this inverted divisor.

---

* This sign $\div$ is really a picture of a fraction, thus emphasizing the fact that the fractional expression is really in itself a symbol of division.

This rule is entirely general, since whole numbers can be written as fractions with a denominator of unity, and mixed numbers reduced to improper fractions before proceeding with the operation. For example:

$$\tfrac{4}{5} : \tfrac{3}{4} = \tfrac{4}{5} \times \tfrac{4}{3} = \tfrac{16}{15}$$
$$\tfrac{3}{4} : \tfrac{5}{6} = \tfrac{3}{4} \times \tfrac{6}{5} = \tfrac{18}{20}$$
$$\tfrac{3}{4} : 5 = \tfrac{3}{4} \times \tfrac{1}{5} = \tfrac{3}{20}$$
$$5 : \tfrac{3}{4} = \tfrac{5}{1} \times \tfrac{4}{3} = \tfrac{20}{3}$$
$$2\tfrac{3}{4} : \tfrac{2}{3} = \tfrac{11}{4} \times \tfrac{3}{2} = \tfrac{33}{8}$$
$$2\tfrac{3}{4} : 3\tfrac{3}{4} = \tfrac{11}{4} \times \tfrac{4}{11} = \tfrac{3}{4}$$

Explanation: If the fractions under consideration possessed the same denominator, it would be unnecessary to take the denominator into consideration at all, and we should only have to divide the numerator of the dividend by the numerator of the divisor. For instance,

$$\tfrac{6}{8} : \tfrac{2}{8} = \tfrac{6}{2} = 3 \,; \; \tfrac{6}{17} : \tfrac{2}{17} = 3 \,; \; \tfrac{6}{35} : \tfrac{2}{35} = 3.$$

Again : $\tfrac{2}{8} : \tfrac{6}{8} = \tfrac{1}{3} \,; \; \tfrac{2}{17} : \tfrac{6}{17} = \tfrac{1}{3}$, etc.

In these cases it is at once apparent that 2 units are contained 3 times in 6 units *of the same kind,* and also that 6 units are contained $\tfrac{1}{3}$ times in 2 units of the same kind.

But this is the same principle as that involved in the above rule for division, since :

$$\frac{6}{8} : \frac{2}{8} = \frac{6}{8} \times \frac{8}{2} = \frac{6}{2} = 3.$$

If now the fractions have not the same denominator, they may be reduced to a common denominator by multiplying numerator and denominator of the dividend by the denominator of the divisor, and also the numerator and denominator of the divisor by the denominator of the dividend. In the example $\tfrac{4}{5} : \tfrac{2}{3}$, these operations would be

$$\frac{4 \times 3}{5 \times 3} : \frac{5 \times 2}{5 \times 3}$$

This gives us the common denominator 5×3 for both fractions, hence it can be omitted, as before, and we simply divide the new numerator (4×3) of the dividend by the new numerator (5 ×2) of the divisor. If we examine this carefully we will see that we have in this operation done exactly what the rule above given directs, namely, inverted the divisor, and then proceeded as in multiplication. Thus:

$$\frac{4}{5} : \frac{2}{3} = \frac{4\times3}{5\times3} : \frac{2\times5}{3\times5} = \frac{4\times3}{5\times2} = \frac{4}{5}\times\frac{3}{2} = 1\frac{2}{10}$$

In the same manner:

$$\frac{5}{7} : \frac{2}{3} = \frac{5\times3}{7\times3} : \frac{7\times2}{7\times3} = \frac{5}{7}\times\frac{3}{2} = 1\frac{1}{14}$$

The following examples are to be worked out by the student according to the rule given at the beginning of this section.

EXAMPLES:

(1) $\frac{4}{16} : \frac{3}{16} = 1\frac{1}{3}$

(2) $\frac{5}{6} : \frac{4}{6} = 1\frac{1}{4}$

(3) $\frac{8}{9} : \frac{2}{3} = 1\frac{1}{3}$

(4) $\frac{7}{4} : \frac{5}{8} = \frac{32}{20}$

(5) $\frac{3}{2\frac{1}{5}} : \frac{1}{3\frac{1}{5}} = 4\frac{1}{5}$

(6) $\frac{1}{3} : \frac{1}{3} = 1$

(7) $\frac{1}{2} : \frac{1}{2} = 1$

(8) $\frac{1}{5} : 5 = \frac{1}{25}$

(9) $1 : \frac{1}{5} = 5$

(10) $\frac{1}{2} : 1 = \frac{1}{2}$

(11) $1 : \frac{1}{2} = 2$

(12) $\frac{5}{17} : 1 = \frac{5}{17}$

(13) $1 : \frac{2}{3} = 1\frac{1}{2}$

(14) $\frac{11}{13} : 7 = \frac{2}{39}$

(15) $1\frac{3}{7} : 17 = \frac{13}{289}$

(16) $3\frac{2}{3} : 7 = \frac{11}{21}$

(17) $11 : 5\frac{7}{9} = 1\frac{47}{52}$

(18) $13\frac{2}{3} : \frac{2}{3} = 20\frac{5}{8}$

(19) $15\frac{1}{2} : 9\frac{7}{8} = 1\frac{5}{8}$

(20) $\frac{2}{3} : \frac{1}{3} = 2$

(21) $\frac{1}{4} : \frac{2}{3} = \frac{3}{8}$

(22) $\frac{4}{5} : 8 = \frac{1}{10}$

(23) $\frac{345}{1000} : \frac{45}{100} = \frac{23}{30}$

(24) $\frac{1}{9} : \frac{1}{10} = 1\frac{1}{9}$

(25) $\frac{1}{10} : \frac{1}{9} = \frac{9}{10}$

(26) $\frac{3}{10} : \frac{17}{100} = 1\frac{13}{17}$

(27) $25\frac{3}{8} : 16\frac{3}{4} = 1\frac{107}{201}$

(28) $\frac{5}{8} : \frac{3}{4} = \frac{5}{6}$

(29) How long is $\frac{1}{3}$ of $\frac{2}{3}$ of a yard? Ans. $\frac{1}{12}$ yard.

(30) How much is $\frac{1}{4}$ of $2\frac{5}{8}$? Ans. $\frac{17}{36}$.

(31) What is the $\frac{1}{4}$ part of $5\frac{4}{5}$ kilograms? Ans. $1\frac{9}{20}$ kg.

(32) How often is $\frac{3}{4}$ contained in 4? Ans. $5\frac{1}{3}$ times.

(33) How many times is $\frac{2}{3}$ contained in 2? Ans. 3 times.

(34) How often is $\frac{1}{3}$ contained in 1? Ans. 3 times.

BOOK V.

Decimal Fractions.

Preliminary Ideas about Approximate Calculations,

48.

In the greater number of cases in which theoretical methods are applied to practical work it is not practicable to obtain results which are absolutely exact, and we must often be content with approximations which are as precise as the conditions will permit. This is always the case when the accuracy of the values from which the data are derived is dependent upon the individual skill of the practical man by whom they have been determined. .

Suppose, for instance, that it is required to measure, by means of a surveyor's chain, the distance between two points about 2000 feet apart. It will be found practically impossible to obtain results which do not vary more or less than half an inch from the true distance. In many cases the personal error, due to the varying ability of different observers, renders it useless to carry calculations beyond a certain closeness to accuracy, and this *personal equation*, as it is called, has been made a subject of calculation itself by means of the Method of Probabilities.

Now in all cases in which it is practically impossible or unnecessary to obtain absolute accuracy, and in which the practical value of the final result will not be impaired if we neglect small fractions, the labor can be greatly reduced and the work simplified by the use of a certain special form of fractions, as will now be explained by an example in addition. It must first be shown that a fraction may be reduced to any denominator we please, without altering its value in the slightest, simply by multiplying both numerator and denominator by the new denominator, and then dividing them both by the old denominator. Thus, for example, the fraction $\frac{6}{8}$ can be converted into one having the denominator 12, as follows : $\frac{6}{8} \times \frac{12}{12} = \frac{72}{8 \times 12}$, and this divided above

and below by 8 gives $\frac{9}{12}$. If the old denominator will not divide the product of the old numerator and the new denominator without remainder, the new numerator will be a mixed number. Thus if we reduce $\frac{5}{8}$ to the denominator 12 we have

$$\frac{5}{8} = \frac{5 \times 12}{8 \times 12} = \frac{7\frac{1}{2}}{12} = \frac{7}{12} + \frac{\frac{1}{2}}{12}$$

Fractions in which the numerator or denominator is a fraction are called *complex* fractions. It is clear that a complex fraction is smaller in value as its denominator becomes greater, thus : $\frac{\frac{1}{2}}{10} < \frac{\frac{1}{2}}{4}$.

If we wish to know how many eighths there are in the fraction $\frac{7}{9}$, we have :

$$\frac{7}{9} = \frac{7 \times 8}{9 \times 8} = \frac{56}{8} = \left(\frac{6\frac{2}{3}}{8} \right)$$

so that we find it to contain 6 whole eighths, that is $\frac{6}{8}$, and something over ; this something being $\frac{2}{3}$ of an 8th, or $\frac{\frac{2}{3}}{8}$

Hence it follows that if we simply annex one, two or three zeros to the numerator of a fraction, and then divide by the denominator, neglecting the fractional remainder, if any, the result will be the equivalent value of the fraction in 10ths, 100ths, or 1000ths, as the case may be.

In the same way, we can find the number of whole tenths, hundredths, or thousandths, etc., which there are in any fraction ; as for example $\frac{7}{8}$, we have :

$$\frac{7}{8} = \frac{\frac{70}{8}}{10} = \left(\frac{8\frac{6}{8}}{10} \right) = \frac{8}{10}$$

$$\frac{7}{8} = \frac{\frac{700}{8}}{100} = \left(\frac{87\frac{1}{2}}{100} \right) = \frac{87}{100}$$

$$\frac{7}{8} = \frac{\frac{7000}{8}}{1000} = \frac{875}{1000}$$

That is, there are 8 *whole* 10ths in $\frac{7}{8}$; or 87 whole 100ths; or 875 whole 1000ths, this last being exact, the two former having a small remainder over.

Let us now attempt to find the sum of the following fractions :

$$\tfrac{1}{2}+\tfrac{1}{33}+\tfrac{4}{5}+\tfrac{6}{7}+\tfrac{7}{9}+\tfrac{5}{11}+\tfrac{9}{13}+\tfrac{13}{247}+\tfrac{3}{6191}+\tfrac{7}{117}$$

these being fractions of same small unit, such as a centimetre, gram, inch, etc.

Now if it is understood that it will be sufficient to obtain the result within $\frac{1}{10}$ of the unit of its exact value, so that the error shall not be more than $\frac{1}{10}$ of a centimetre, gramme, inch, etc., more or less, the labor can be very much reduced in the following manner. Instead of reducing the fractions to a common denominator (which, according to the method explained in § 30, would be a very large number), we take for the common denominator one of the simple unit numbers—10, 100, 1000, etc., whichever is the most convenient. Suppose that we take 100 for the common denominator, and by the method just explained convert the given fractions by (mentally) adding two zeros to each numerator and dividing by the denominators, and if the fractional remainder is less than $\frac{1}{2}$ it may be neglected, while if it is greater than $\frac{1}{2}$ it may be called 1. We thus get the following values, all understood to be hundredths.

$$
\begin{aligned}
&\text{100ths}\\
\hline
\tfrac{1}{2} &= 50\\
\tfrac{1}{33} &= 33+\tfrac{1}{33}\\
\tfrac{4}{5} &= 80\\
\tfrac{6}{7} &= 86-\tfrac{2}{7}\\
\tfrac{7}{9} &= 78-\tfrac{2}{9}\\
\tfrac{5}{11} &= 45+\tfrac{5}{11}\\
\tfrac{9}{13} &= 69+\tfrac{3}{13}\\
\tfrac{13}{247} &= 5+\tfrac{65}{247}\\
\tfrac{3}{6191} &= 0+\tfrac{300}{6191}\\
\tfrac{7}{117} &= 6-\tfrac{18}{117}
\end{aligned}
$$

Total, $\frac{452}{100}=4\frac{52}{100}$

Explanation : If, in this approximate calculation, we had neglected $\frac{1}{2}$ of a fractional unit $\dfrac{\frac{1}{2}}{100}$ for each number, the total error in the sum of the ten figures could not have exceeded $\frac{5}{100} = \frac{\frac{1}{2}}{10}$.

But we have only neglected fractions of less than $\frac{1}{2}$ a unit, and as these have been taken sometimes on one side and sometimes on the other, it follows that the error in the sum $4\frac{11}{66}$ must be less than $\frac{1}{2}$.

If the common denominator had been taken as 100000 or 1,000,000 instead of 100 the possible error would have been enormously reduced, and become practically inconceivably small even though the unit itself be as large as a kilogram[or a mile.

49.

Common fractions require, as we know, two numbers to be used in writing them, a numerator and a denominator, and this is more or less inconvenient, especially in tabular matter, where room is a question of importance. The form of common fractions is also a matter of inconvenience in various ways, such as the necessity of reduction to a common denominator for purposes of addition or subtraction, and hence the invention and introduction of decimal fractions which do away with all this inconvenience, is an improvement of the highest importance, especially for long and intricate computations.

50.

Decimal Fractions, are those fractions of which the denominators are any one of the units of the decimal system of notation, as for example, $\frac{873}{1000}$, $\frac{47}{100}$, $\frac{3}{10}$, $\frac{3}{1000}$, etc.

Such a fraction, as a necessary consequence of its form, has the property of being subdivided into as many separate fractions as it has zeros in the denominator, and the denominators of these separate fractions follow each other in the regular decimal order, thus, $\frac{1}{10}$, $\frac{1}{100}$, $\frac{1}{1000}$. For example, the number $873 = 800 + 70 + 3$, and likewise the fraction

$$\frac{873}{1000} = \frac{800}{1000} + \frac{70}{1000} + \frac{3}{1000}$$

or by canceling those zeros which appear both in numerator and denominator:

$$\frac{873}{1000} = \frac{8}{10} + \frac{7}{100} + \frac{3}{1000}$$

In like manner we have

$$\frac{34507}{100000}=\frac{3}{10}+\frac{4}{100}+\frac{5}{1000}+\frac{0}{10000}+\frac{7}{100000};$$

$$\frac{47}{1000}=\frac{0}{10}+\frac{4}{100}+\frac{7}{1000};$$

$$\frac{47}{100000}=\frac{0}{10}+\frac{0}{100}+\frac{0}{1000}+\frac{4}{10000}+\frac{7}{100000}.$$

in which, for sake of uniformity, the numerators which are lacking have been replaced by zeros.

51.

From the above analysis we see that the numerator of a decimal fraction has as many figures as there are zeros in the denominator, or else zeros can be prefixed to the numerator to fill the deficiency.

Thus $\frac{875}{1000}$, $\frac{71}{100}$ are examples of the first case, and $\frac{47}{1000}=\frac{047}{1000}$, $\frac{47}{10000}=\frac{0047}{10000}$ illustrate the second case. It also appears that the first figure of the numerator shows the number of tenths, the second the number of hundredths, the third the number of thousandths, etc.

These features may be utilized to enable us to write decimal fractions in a very simple manner without the necessity for writing the denominator at all. This accomplished by the use of the *decimal point*, the figures of the numerator being written to the *right* of a period (.), there being always as many figures written as there are zeros in the denominator of the decimal fraction.* If the numerator does not contain as many figures as there are zeros in the denominator, then after writing the decimal point, zeros are used to fill in the deficiency *before* the figures are written, always writing to the *right* from the decimal point.

If there are any whole numbers to be written with the decimal, these are written to the left of the decimal point.

Thus, for example, $\frac{875}{1000}=0.875$, which is read o units, 8 tenths, 7 hundredths, and 5 thousandths; or, more simply, 875 thousandths. $\frac{47}{10000}=0.0047$, which is read o units, o tenths, o hundredths, 4 thousandths, 7 ten thousandths; or, more simply, 47 ten-thousandths.

In the same way we may write:

$$5\frac{47}{1000}=5.047; \quad 24\frac{805}{1000}=24.805; \quad 32\frac{5}{100}=32.05.$$

---

*In Germany, and on the continent generally, a comma (,) is used instead of a period, and care must be taken in reading mathematical works to avoid confusion from this source. The period is invariably used in England and America.

We see, therefore, that it is easy to write decimal fractions without the use of a denominator, and the reverse is also true, that any decimal fraction can at once be written with a denominator, as the denominator will always consist of unity followed by as many zeros as there are places of figures following the decimal point. Thus we have: $0.875=\frac{875}{1000}$; $0.0071=\frac{71}{10000}$.

EXAMPLES. Write the following fractions as decimals:

$\frac{5}{10}=0.5$ $\qquad$ $\frac{376}{10000}=0.0376$

$\frac{1}{10}=0.1$ $\qquad$ $\frac{2005}{100000}=0.02005$

$\frac{13}{100}=0.13$ $\qquad$ $\frac{501007}{1000000}=0.501007$

$\frac{3}{100}=0.03$ $\qquad$ $3\frac{2}{10}=3.2$

$\frac{101}{10000}=0.0101$ $\qquad$ $728\frac{47}{100}=728.47$

$\frac{1}{10000}=0.0001$ $\qquad$ $10\frac{15}{1000}=10.015$

$\frac{3}{1000}=0.003$ $\qquad$ $11\frac{1101}{100000}=11.01101$

$\frac{75}{100}=0.75$ $\qquad$ $\frac{50013}{1000}=50.013$

EXAMPLES. Convert the following decimals into common fractions:

$0.54=\frac{54}{100}$ $\qquad$ $30.07=30\frac{7}{100}=\frac{3007}{100}$

$0.015=\frac{15}{1000}$ $\qquad$ $0.005=\frac{5}{1000}$

$2.004=2\frac{4}{1000}=\frac{2004}{1000}$ $\qquad$ $100.001=100\frac{1}{1000}=\frac{100001}{1000}$

Conversion of Common Fractions into Decimal Fractions.

52.

Decimal fractions are found extremely useful in applied mathematics, especially in connection with the extraction of roots, and with the use of logarithms, as will be seen hereafter. Logarithms could hardly be used without decimals, and, indeed, it was the use of decimals which led to the introduction of logarithms. Although decimals are most frequently used directly in their own form in actual practice, yet it is often required to convert a common fraction into the decimal form, and this is most readily done in the following manner: Write o for the units place, also write the decimal point; annex a zero to the numerator, and divide by the denominator, the quotient will give the

number of tenths which the fraction contains, which write next
on the right of the decimal point (if there are no tenths, write a
zero in the place); annex a zero to the remainder and divide
again by the denominator, the quotient will be hundredths; proceed thus with the division until as many decimal places have
been determined as the precision of the calculation demands.
More than seven places are rarely required; in many cases two,
three or five are sufficient. The last decimal figure should be increased by 1 when the remainder after the last division exceeds 5.
For example, if the decimal 0.8468 is to be used only to three
places, the value will be more accurately given by 0.847 than by
0.846; the former is too large by only $\frac{2}{10000}$, while the latter is
too small by $\frac{8}{10000}$.

If the denominator of a fraction can be separated into the
simple factors 2 and 5, as $\frac{3}{8} = \frac{3}{2 \times 2 \times 2}$, or $\frac{3}{40} = \frac{3}{2 \times 2 \times 5}$, the division can be carried out exactly, but if the denominator cannot be
separated into the simple factors 2 and 5, as $\frac{5}{6} = \frac{5}{2 \times 3}$, or $\frac{4}{15} =$

$\frac{4}{3 \times 5}$, the division can never be exactly completed without a
remainder.   See § 317.

Examples :

$\frac{3}{16} = 0.1875$, exactly.

$\frac{7}{8} = 0.875$, exactly.

$\frac{6}{2117} = 0.0028 \ldots \ldots$

The actual work on the first of the above examples is as
follows :

$$
\begin{array}{r}
16 \,)\; 3.0 \quad (\; 0.1875 \\
\underline{1\;6} \\
1\;40 \\
\underline{1\;28} \\
1\,2\,0 \\
\underline{1\,1\,2} \\
8\,0 \\
8\,0
\end{array}
$$

This, expressed in words, would be: 16 into 3, no times;
16 into 30 tenths gives 1 tenth and 14 remainder; 16 into 140

hundredths gives 8 hundredths and 12 remainder; 16 into 120 thousandths gives 7 thousandths and 8 remainder; 16 into 80 ten-thousandths gives 5 ten-thousandths.

The principle of the above method for converting a proper fraction into a decimal is easily seen, for if, (according to §48) the numerator and denominator of the fraction $\frac{3}{16}$ are first multiplied by the new denominator (in this case 10,000) and then divided by the old denominator (16) we have practically the same operation, thus: $\frac{3}{16} = \frac{30000}{16 \times 10000} = \frac{1875}{10000} = 0.1875$

If the common fraction is an improper fraction, the whole portion must, of course, be written before the decimal point, instead of the 0, as directed above, thus:

$$\frac{21}{8} = 2.625 ; \quad 2\frac{3}{4} = 2.75, \text{ etc.}$$

If any number of zeros be annexed at the *right* of a decimal the value will thereby not be altered, there being no more effect than if zeros were placed at the *left* of a whole number. For example: 0.54=0.540=0.5400, etc.

Zeros are sometimes thus annexed merely to secure uniformity among a number of decimal fractions by making them all of the same denomination. When an exact division cannot be made in reducing a common fraction to a decimal, it will be found that after a certain time the figures will be repeated in the same order as before. Such fractions are termed "periodical" or "repeating" decimals, and the return of the period indicated by dots, thus: $\frac{5}{99} = 0.0505 \ldots$, with the period 05; $\frac{2}{3} = 0.0666 \ldots$, with the period 6.

EXAMPLES. Convert the following fractions into decimals. If the division cannot be made exactly, five decimal places will be sufficient.

$\frac{3}{4} =$	0.75	$3\frac{700}{301} =$	12.29236
$\frac{9}{25} =$	0.36	$\frac{1595}{245} =$	6.5102
$\frac{5}{7} =$	0.71429	$\frac{300}{1157} =$	0.25929
$\frac{1}{9} =$	0.11111 . . .	$20\frac{3}{8} =$	20.375
$\frac{2}{9} =$	0.22222 . . .	$304\frac{5}{7} =$	304.71429
$\frac{4}{9} =$	0.44444 . . .	$\frac{1}{93} =$	0.01075
$\frac{1}{2} =$	0.5	$\frac{1}{10} =$	0.1
$\frac{2}{5} =$	0.4	$\frac{3}{1000} =$	0.003

$\frac{100}{3374}=$ 0.02964 $\frac{37}{36000}=$ 0.00103

$\frac{1}{99}=$ 0.010101 . . . $\frac{141}{40000}=$ 0.00352

$\frac{11}{474}=$ 0.02321 $\frac{7}{40}=$ 0.175

$\frac{1}{8}=$ 0.125 $\frac{3}{200}=$ 0.015

                                              $\frac{77}{600}=$ 0.12833 . . .

See Appendix § 316.

53.

The decimal system is especially convenient when we consider a whole number in combination with a decimal fraction, as the same system enables us to pass from the fractional units to the whole ones. Thus in the number:

$$20704.56803$$

we see that ten units of the second decimal place (6) equal one unit of the first decimal place (5), since $10 \times \frac{1}{100} = \frac{1}{10}$; ten units of the first decimal place equal one of the whole units place (4), etc. For this reason all calculations with decimals are performed in exactly the same manner as with whole numbers, it being only necessary to exercise a little care with regard to the position of the decimal point.

Addition.

54.

The quantities to be added are under each other; care being taken to place units of the same kind under each other, units under units, tenths under tenths, etc. The addition is performed just the same as with whole numbers; for every ten units in any column, one unit is carried to the next higher column. If any of the quantities contain common fractions they must be reduced to decimal fractions before the addition is commenced. For example:

(1) $0.72 + 0.087 + 2.5 + 14.0089 =$

           0.72
           0.087
           2.5
           14.0089
           ――――――
           17.3159   Answer.

(2) $0.05012+30.0707--+0.66---+\frac{5}{6}+2\frac{3}{4}=$

$$
\begin{array}{r}
0.050120 \\
30.070707 \\
0.666667 \\
\tfrac{5}{6}= 0.833333 \\
2\tfrac{3}{4}= 2.750000 \\
\hline
34.370827\text{—Answer.}
\end{array}
$$

(3) $10.3131--+9.11--+0.503\times0.003+0.1=$

$$
\begin{array}{r}
10.313131 \\
9.111111 \\
0.503000 \\
0.003000 \\
0.100000 \\
\hline
20.030242
\end{array}
$$

Subtraction.

55.

The subtrahend is written under the minuend in the same manner as with whole numbers, and the subtraction made as usual. Common fractions must first be converted into decimals.

Examples :

	(1)	(2)
Minuend	1.0407	8.000
Subtrahend	0.9745	7.995
Difference	0.0662	0.005
	(3)	(4)
Minuend	13.66667	$\frac{3}{4}=0.7500$
Subtrahend	3.67809	0.2305
Difference	9.98858	0.5195

Multiplication.

56.

Rule 1. When a decimal is to be multiplied by a decimal unit (such as 10, 100, 1000, &c.), it is only necessary to move the decimal point as many places to the right as there are zeros in the multiplier. For example : $10 \times 0.045 = 0.45$; $100 \times 0.045 =$ 4.5 ; $1000 \times 0.045 = 45$; $10000 \times 0.045 = 450$; $100 \times 2.003 = 200.3$.

RULE 2. In order to multiply two decimals together, proceed with the multiplication exactly as with whole numbers, paying no attention to the decimal point; then point off as many decimal places in the product as there are in *both the factors together*. If the product has fewer places than this requires, supply the deficiency with zeros.

EXAMPLES :

(1)	0.43	(2)	8.034	(3)	0.0478
	0.25		0.46		0.003
	215		48204		0.0001434
	86		32136		
	0.1075		3.69564		

(4)	4.03	(5)	0.035	(6)	0.056
	2.15		2.04		24
	2015		140		224
	403		70		112
	806		0.07140̸		1.344
	8.6645				

In example (1) both factors together have four decimal places ; in (2) five, in (3) seven, in (5) four, in (5) five, in (6) three.

The reason for this rule will be apparent if we consider the two factors as fractions with their denominators written under them, multiplying numerator by numerator and denominator by denominator, and then finally write the product as a decimal again, without a denominator. Thus :

$$0.43 \times 0.25 = \tfrac{43}{100} \times \tfrac{25}{100} = \tfrac{1075}{10000} = 0.1075$$

and it is seen at once that the denominator of the product contains as many zeros as there are in the denominators of both factors together, which determines the number of decimal places in the answer. In the same manner we have :

$$8.034 \times 0.46 = \tfrac{8034}{1000} \times \tfrac{46}{100} = \tfrac{369564}{100000} = 3.69564.$$

$$0.056 \times 24 = \tfrac{56}{1000} \times 24 = \tfrac{1344}{1000} = 1.344 \quad \text{(see § 51).}$$

EXAMPLES :

(7) $0.057 \times 0.005 = 0.000285$

(8) $0.205 \times 7.04 = 1.4432$

(9) $1.09 \times 1.003 = 1.09327$

(10) $11 \times 1.1036 = 12.1396$

(11) $0.013 \times 101 = 1.313$

(12) $203.07 \times 105.002 = 21322.75614$

(13) $100 \times 0.031 = 3.1$

(14) $0.2 \times 100 = 20$

(15) $1000 \times 31.0451 = 31045.1$

(16) $21.005 \times 0.74 \times 0.07 = 1.0880590$

The above examples are merely stated with their answers but should be fully worked out in detail by the student.

Division.

57

RULE 1. If a decimal is to be divided by a decimal unit (10, 100, 1,000, &c.) it is only necessary to move the decimal point to the left, as many places as there are zeros in the divisor. For example :

$$\frac{320.45}{100} = 3.2045$$

$$\frac{5.23}{10} = 0.523$$

$$\frac{1.04}{100} = 0.0104$$

RULE 2. If the divisor is a whole number, the dividend is divided directly by it, continuing the division as far as may be desired, instead of setting down the usual remainder. Thus :

$$\frac{3.645}{8} = 0.455625$$

This is as if we said : the 8th part of three units is no unit ; the 8th part of 36 tenths is 4 tenths, and 4 remainder ; the 8th part of 44 hundredths is 5 hundredths, and 4 remainder, &c., &c. In like manner we have :

$$\frac{0.06305}{12} = 0.005254$$

Rule 3. If the divisor is a decimal, proceed as follows : Move the decimal point in the divisor as many places to the right as are necessary to convert it into a whole number. ; Also move the decimal point in the dividend to the right as many places as has been done in the divisor ; then proceed as in rule 2.

For example :

$$\frac{0.057}{3.2}=?$$

Here the divisor has one decimal place, so the decimal point must be moved to the right one place in both numbers, and we have

$$\frac{0.57}{32}=0.01781$$

This follows from the fact that we can multiply both numerator and denominator of a fraction by any number without changing its value and the above operation is practically as follows :

$$\frac{0.057}{3.2}=\frac{0.057\times 10}{3.2\times 10}=\frac{0.57}{32}=0.01781$$

Again :

$$\frac{0.35}{0.4073}=?$$

Here the divisor has four decimal places, and hence the decimal point must be moved four places to the right in both numbers, and we have

$$\frac{0.35}{0.4073}=\frac{3500}{4073}=0.85932$$

The following are additional examples of the same principle :

$$\frac{0.34}{0.17}=\frac{34}{17}=2$$

because

$$\frac{0.34}{0.17}=\frac{0.34\times 100}{0.17\times 100}=\frac{34}{17}=2$$

$$\frac{8}{0.245}=\frac{8\times 1000}{0.245\times 1000}=\frac{8000}{245}=32.653064$$

$$\frac{0.3645}{0.8}=\frac{0.3645\times 10}{0.8\times 10}=\frac{3.645}{8}=0.455625$$

When a decimal fraction is to be multiplied by a common fraction, the latter may first be converted into a decimal, but it is usually simpler to multiply the decimal by the numerator of the common fraction, and divide the product by the denominator.

Thus :
$$\tfrac{2}{3} \times 6.0435 = \frac{2 \times 6.0435}{3} =$$
$$= 2 \times 2.0145 = 4.029$$

$$\tfrac{3}{4} \times 0.25 = \frac{3 \times 0.25}{4} = \frac{0.75}{4} = 0.1875$$

If a decimal fraction is to be divided by a common fraction, we have simply to multiply it by the fraction inverted. Thus :

$$0.326 \div \tfrac{3}{4} = 0.326 \times \tfrac{4}{3} = \frac{4 \times 0.326}{3} =$$
$$= \frac{1.304}{3} = 0.43466$$

If a common fraction is to be divided by a decimal we must first convert the common fraction into a decimal, or else we may convert the decimal into a common fraction and then invert it and multiply thus :

$$\frac{3}{4} : 0.321 = \frac{0.75}{0.321} = \frac{750}{321} = 2.3364$$

or else :

$$\frac{3}{4} : \frac{321}{1000} = \frac{3}{4} \times \frac{1000}{321} = \frac{250}{107} = 2.3364$$

EXAMPLES: Perform the divisions indicated in the following examples, carrying the results out to four places of decimals :

(1) $\dfrac{0.305}{0.506} = 0.6028$

(2) $\dfrac{0.0578}{0.23} = 0.2513$

(3) $\dfrac{2.035}{0.76} = 2.6776$

(4) $\dfrac{0.03}{8.134} = 0.0037$

(5) $\dfrac{3.207}{4.085} = 0.785$

(6) $\dfrac{0.091}{183} = 0.0005$

(7) $\dfrac{25}{0.507} = 49.3097$

(8) $\dfrac{0.0453}{8} = 0.0056$

(9) $\dfrac{2.1616 \ldots}{12} = 0.1801$

(10) $\dfrac{0.04}{10} = 0.004$

(11) $\dfrac{540.047}{1000}$=0.540047   (12) $\dfrac{0.04}{100}$=0.0004

(13) $\dfrac{8471}{100}$=84.71   (14) $\dfrac{1}{100}$=0.01

(15) $\dfrac{35}{1000}$=0.035   (16) $\dfrac{1}{0.305}$=3.2787

(17) $\dfrac{1}{1.012}$=0.9881   (18) $\dfrac{2}{3.05}$=0.6557

(19) $\frac{2}{7} \div 0.14$=2.0408   (20) $4.03 \div \frac{2}{3}$=6.045

(21) $0.1875 \div \frac{3}{4}$=0.25   (22) $0.056 \div \frac{3}{7}$=0.1307

(23) $0.435 \div 2\frac{3}{4}$=0.1582   (24) $\frac{3}{8} \div 0.05$=7.5

(25) $\dfrac{0.03}{0.05} \times \frac{2}{3}$=0.4

BOOK VI.

Calculating with Denominate Numbers.

After having learned how to calculate with abstract numbers it is easy to reckon with denominate quantities, as the only thing necessary is to learn the different kinds of units in use, as well as their sub-divisions; that is to say, the various systems of weights, measures, and monetary values.

Addition.

58.

When magnitudes of different denominations are to be added together, all quantities of the same denomination are set under each other, and each column added, beginning with the lowest denomination. The sum of each column, as added, is reduced to that of the next higher denomination.

Examples:

(1)

Dollars	Cents
8	45
11	32
126	89
0	28
17	2
163	96

(2)

Kilograms
27.537
29.756
37.825
20.762
8.059
123.939 kg.

(3)

Hours	Minutes	Seconds
5	37	18
18	48	16
9	20	39
12	11	42
5	7	8
51 h.	5 min.	3 sec.

(4)

Feet	Inches
12	7
23	5
9	11
27	2
84	3
157 ft.	4 in.

Subtraction.

59.

The quantities are placed as in addition, that is, those of like denomination under each other. If any quantity in the subtrahend is greater than that in the same denomination of the minuend, one unit is taken from the next higher denomination of the minuend and reduced to the next lower denomination.

Thus :

	Hours	Minutes	Seconds
(1)	10	45	30
	7	59	45
	2 h.	45 m.	45 sec.

The preceding examples are all in such well-known denominations that no especial explanation is necessary. There is, however, another case relating to the subtraction of one quantity of time from another, which will here be explained.

When it is desired to find the difference between two quantities of time, both of which are reckoned from the same starting point, that is, from the year of the birth of Christ, and hence expressed in the form of dates, a somewhat different procedure is necessary. We must remember that the civil day begins at 12 o'clock midnight, and lasts for $2 \times 12 = 24$ hours. We also have the times given in dates, but require the answer to tell the number of years, months, days and hours which have elapsed between them.

Hence, for instance, at 7 o'clock in the evening, on April 25, 1834, there had elapsed from the Christian era, 1833 whole years, 3 whole months, 24 whole days, and $12 + 7 = 19$ hours. The number of days in the various months must be memorized, and this task can be made easy by arbitrary rules. February is the exception, having 28 days, except in leap year when it has 29. (A leap year is one in which the date can be divided by 4 without a remainder.) If we repeat the names of the months over, starting with January, and counting on the knuckles and the spaces between the knuckles, every month which falls on a knuckle will contain 31 days, and those which fall between the knuckles will have 30 days, excepting February.*

We will now illustrate by an example :

(1) How much time has elapsed between 5.31 P. M., October 24, 1790, and 11.27 A. M., March 22, 1832 ?

OPERATION :

	Years	Months	Days	Hours	Min.
			29		
1832, March 22, 11 h., 27 m., A. M.=1831		2	21	11	27
1790, October 24, 5 h., 31 m., P. M.=1789		9	23	17	31
Difference	41	4	26	17	56

Note that 1832 is a leap year, and hence the month preceding March is February, 29 days, and hence 29 days are added to 21 when a month is borrowed from the column of months.

(2) What will be the date when 60 years, 2 months, 16 days, 17 hours and 50 minutes have elapsed since 11.25 P. M., July 17, 1819?

	Years	Months	Days	Hours	Min.
The given date is	1818	6	16	23	25
Adding	60	2	16	17	50
	1878	9	3	17	15

Hence the date is 5.15 P. M., October 4, 1879.

Multiplication.

60.

The multiplier must itself always be an abstract number, since no denominate number can be multiplied by another denominate number. If the multiplicand is of several denominations, each of these is multiplied by the multiplier separately, and the product reduced to the next higher unit.

We must be careful to observe that what may at first appear to be the multiplying of one denominate number by another is really only apparently so.

---

*This is an old German rule. In England and America everyone goes by the old jingle :
" Thirty days hath September,
April, June and November.
All the rest have thirty-one,
Excepting February alone,
To which we twenty-eight assign
'Till Leap Year gives it twenty-nine."

Suppose we ask how much 7 pounds of a given article would cost at $6.72 a pound—we have :

$$\begin{array}{r} \$6.72 \\ 7 \\ \hline \$47.04 \end{array}$$

Here we have not multiplied dollars by pounds, although it may look so. One pound costs $6.72, hence 7 pounds cost 7 *times* as much, and the multiplier 7 is an *abstract* and not a denominate number.

EXAMPLES. (1) A meter contains 100 centimeters. How much is 5$\frac{4}{7}$ times 7 meters, 83 centimeters? Ans. 44 m., 74 cm.

(2) A kilogram contains 1000 grams. How much is 2$\frac{3}{4}$ times 2 kilograms, 430 grams? Ans. 6 kg., 682.5 grms.

(3) Multiply 3 hours, 26 minutes, 12 seconds by $\frac{5}{6}$.
Ans. 2 hrs., 51 min., 50 sec.

Division.

61.

First Case. If it is required to divide the dividend into a given number of parts, the divisor is always an abstract number, while the quotient, being a part of the dividend, is of the same denomination. The highest units are divided first, and the remainder after each division reduced to the next lower unit. For example : Required the 8th part of 29 days, 15 hours, 14 minutes. We have :

	Days	Hours	Min.
8)	29	15	14

3 d. 16 hrs. 54$\frac{1}{4}$ min.

Second Case. If both divisor and dividend are denominate quantities, they must both be reduced to units of like denomination, whether higher or lower, since division can only be performed with quantities of the same denomination. The quotient in this case is always an abstract number, and simply indicates the number of times the one quantity is contained in the other.

For example :

How many times is 3 hours, 20 minutes contained in 11 hours, 40 minutes?

11 hours, 40 minutes=700 minutes.

3 hours, 20 minutes=200 minutes.

$\frac{700}{200} = 3\frac{1}{2}$ times.

EXAMPLES :

(1) What is the 24th part of 130 dollars and 20 cents?

Ans. $5.42½.

(2) How many times greater is 8 metres, 11 centimetres, than 2 metres, 6 centimetres? Ans. $3\frac{199}{206}$.

BOOK VII.

Direct, Inverse and Compound Proportion. (The Rule of Three.)

· Direct Proportion.

62.

When two quantities bear such a relation to each other that when either one of them alters its value in the slightest degree, becoming either greater or smaller, the other becomes also proportionally greater or smaller, the two quantities are said to be in *Direct Proportion* to each other. Thus, for example, a number of articles and their value are in direct proportion to each other since it is evident that the value of two or three times the number would be two or three times as great as that of the original number, or that $\frac{1}{3}$ the original number would be worth $\frac{1}{3}$ the value.

Examples of this sort, in which the variation of one magnitude is given and the corresponding variation is required of some magnitude which is dependent upon the given one, are of much importance and frequent occurrence in every day life. The principle of stating and solving such problems is always the same, and when one understands the simplest examples, he can readily solve any which may occur. EXAMPLE. If 5 yards of cloth cost 54 shillings, what would be the cost (in the same proportion) of 100 yards?

First Solution. As we must pay 54 shillings for every 5 yards, we may consider 5 yards as a unit of measurement, and hence shall have to pay as many times 54 shillings as 5 is contained in the given number of yards. Therefore we have

$$\frac{100 \text{ yds.}}{5 \text{ yds.}} \times 54 \text{ shillings} = \frac{100}{5} \times 54 = 20 \times 54 = 1080 \text{ s.}$$

Second Solution. We may reduce the given value first to the actual unit of measurement, and say : If 5 yards cost 54 shillings

1 yard will cost $\frac{1}{5}$ as much, that is $\frac{54}{5}$ shillings, and 100 yards will cost 100 times as much, whence:

$$100 \times \tfrac{54}{5} = 1080 \text{ shillings (see § 60).}$$

Either of the above solutions may be readily stated in the form of a rule of Direct Proportion, or the so-called "Rule of Three."*

We may say: If 5 yards cost 54 s., what will be the cost of 100 yards? and multiply the second and third quantities together and divide by the first, considering the first and third quantities, both having the same name, as undenominate numbers.

The operation may very often be simplified by factoring the numbers and canceling out those factors which appear in both dividend and divisor.

If the first and third quantities are not of the same denomination they must first be reduced either to the next lower unit or to the highest unit; the latter often involving the least labor, since the smaller units may at once be put in the form of a fraction of the higher. (See § 61, 62.)

When two quantities stand in *Direct Proportion* to each other we may avoid mistakes by saying mentally "the greater, the greater," or "the less, the less," and thus prevent confusion with Inverse Proportion, which will be explained hereafter. Thus in the above example, 100 yards is *greater* than 5 yards, and 1080 shillings is *greater* than 54 shillings. If the question had been: 100 yards cost 1080 shillings, how much would 5 yards cost? we should have had 5 yards is less than 100 yards, and so 54 shillings is less than 1080 shillings. It must therefore be remembered that in *Direct* Proportion both quantities either increase or diminish together.

EXAMPLES:

(1) If \$100 earns \$5 interest in one year, how much will \$625 earn in one year?

Answer. $\frac{625}{100} \times 5 = \$31\frac{1}{4}$ interest; i. e., "the greater" capital "the greater" interest, hence 5 will be contained in the answer as often as 100 is contained in 625; or, if 100 give 5, how much will 625 give?

(2) What will be the interest on \$1065.25 for one year at 5 per cent.? (By per cent. is meant simply the interest on every

---

* Called the "Rule of Three" because three quantities are given and the fourth required.

hundred, i. e., at a rate of 5 per cent. the interest is 5 dollars for every hundred dollars, 5 cents for every 100 cents, etc., etc.)

Answer. $53.26¼.

(3) If $14,560 earns $364 interest in one year, what is the per cent.? That is, what interest would $100 earn?

Answer. $\frac{100}{14560}\times 364 = 2\frac{1}{2}\%$.

The symbol % is used to indicate per cent.

In the calculation of interest a certain *time* is always given, i. e., 1 year, ½ year, 1 month, etc. If, therefore, the time is greater or less than the unit for which the *rate* of interest is stated, the *time-unit* must be taken as many times, or parts of times, as the question states. For example:

(4) What will be the interest on $600 in *3 years*, if it earns 4% *a year*? i. e. if the *rate* is 4% *a year*.

Answer. $\frac{600}{100}\times 4\times 3 = \$72$. We multiply by 3 because it would earn 3 times as much in 3 years as it does in 1 year.

(5) What is the interest on $34.50 in 6 years, 3 months at $2\frac{1}{3}\%$?

Answer. $\frac{34.50}{100}\times 2\frac{1}{3}\times 6\frac{1}{4} =$

$$= \frac{69}{2\times 100}\times \frac{7}{3}\times \frac{25}{4} = \frac{23\times 7}{2\times 4\times 4} = \frac{161}{32} = \$5\frac{1}{32}$$

or, converting $\frac{1}{32}$ into a decimal $= \$5.03$

(5) What will be the interest on $73.75 for 14 days at 5% a year?

Answer. $\frac{73.75}{100}\times 5\times \frac{14}{365} = 14.144$ cents, or 14 cents.

When a bill is to be discounted the discount is a certain percentage calculated for the length of time which will elapse before the bill falls due, this amount, or discount, being deducted from the value of the face of the bill, gives its present value.

EXAMPLE:

(6) A draft for $600 at 4 weeks, is to be discounted at the rate of 8 per cent. (a year), how much will the discount be?

Answer. $\frac{600}{100}\times 8\times \frac{4}{52} = \frac{6\times 8}{13} = \$3\frac{9}{13}$.

In countries where the money is based on the decimal system it is best to make all interest calculations in the decimal system also, as this admits of the use of very simple methods, and avoids the introduction of awkward fractions in the final result.

The best plan is to reduce the time to months and decimals of a month, and, as the legal month consists of 30 days, we can always reduce days to tenths of a month by dividing by 3. Thus, 5 months, 18 days=5.6 months; 2 years, 5 months, 18 days=29.6 months.

The rate for 1 month is $\frac{1}{12}$ of the rate for 1 year, hence we have only to divide the capital by 100 and by 12, and multiply by the rate and the time (in months and decimals) to obtain the interest.

Example :

(7) What is the interest on \$1252.75 at 6 per cent. for 3 years, 6 months, 12 days?

Answer.   3 yrs., 6 mo., 12 d.=42.4 mo.

hence, $\frac{1252.75}{100} \times \frac{6}{12} \times 42.4 = \$265.58.$

In business and commercial establishments, where calculations of interest and percentages are constantly required, interest tables are used, such tables having been computed in a very complete and exhaustive manner.

Inverse Proportion.

63.

When two quantities bear such a relation to each other that when either one of them alters its value in the slightest degree, becoming either greater or smaller, the other becomes inversely smaller or greater, the two quantities are said to be in *Inverse Proportion* to each other.

It must always be remembered that in *inverse* proportion the two quantities vary in the *opposite* direction, one becoming smaller as the other becomes greater, or greater as the other becomes smaller. This can be memorized by saying mentally :, "the more—the less," or "the less—the more."

If, for example, 4 workmen can accomplish a given piece of work in 6 days, twice as many workmen, that is 8 men, will do the work, not in *twice* the time, but inversely, in *one-half* the time; hence "the more" the men, "the less" the time, and therefore we say that the number of men and the amount of time stand in *inverse* proportion to each other. In like manner, if 4 men can accomplish a piece of work in 6 days, and it is asked how many men will be required to accomplish it in 2 days, we see clearly that the number of men must be *increased* in the same proportion as the time is *decreased*. We must learn in all examples of this sort, to remember that the quantity about which the question is asked, is to be divided by the quantity which is given (this latter being regarded as a scale or unit of measurement) and then use the quotient to divide the quantity whose magnitude has changed, or (what is the same thing) multiply by the quotient *inverted*.

Stating this as a Rule of Three, we must multiply the first two together and divide by the third, and as this is the inverse of the rule for direct proportion, it may be called the Inverse Rule of Three.

EXAMPLES :

(1) 6 men can perform a piece of work in 3 days, how many men will be required to do it in 2 days?

Answer. $\frac{3}{2} \times 6 = 9$

or, If 3 days require 6 men, how many will be required for 2 days?

$$3 \times 6 = 18$$
$$18 \div 2 = 9 \text{ men.}$$

(2) 6 men can perform a piece of work in 7 hours, in how many hours can it be done by eight men?

Answer. $\frac{6}{8} \times 7 = 5\frac{1}{4}$ hours. "The more" men "the less" time.

(3) It requires $3\frac{1}{2}$ yards of cloth $1\frac{1}{4}$ yards wide to make a coat, how many yards will it take if the cloth is $\frac{3}{4}$ yards wide? "*The less*" the width, "*the more*" the length.

$$\frac{\frac{5}{4}}{\frac{3}{4}} \times 3\frac{1}{2} = \frac{5}{3} \times \frac{7}{2} = \frac{35}{6} = 5.833 \text{ yds.}$$

(4) There are in a fortress 600 men, supplied with a quantity of bread which will last them 4 months at 2 pounds a day for each man. If 400 more men are placed in the fortress, and the same amount of bread is required to last 4 months, how much will the daily allowance be for each man?

Answer. $\frac{600}{1000} \times 2$ lbs. $= 1.2$ lbs.

Compound Proportion.

64.

A quantity may depend upon several other quantities in such a manner that it may be either in direct or inverse proportion with each of them, and at the same time be in *compound* proportion with all of them together. For example, if 6 stone masons can in 7 days build a wall 4 stones thick, 3 feet high and 120 feet long, and it is required to know how long a time would be required for 12 stone masons, working at the same known rate, to build a wall 2 stones thick, 9 feet high and 60 feet long, we can solve this simple problem, or any similar one in the following manner:

The question is stated by writing the known case under the unknown, in such a manner that like quantities shall be under each other, thus :

| 12 men | ? time | 2 stones thick | 9 ft. high | 60 ft. long | (unknown case) |
| 6 men | 7 days | 4 stones thick | 3 ft. high | 120 ft. long | (known case) |

Here we have the unknown quantity (time) in the upper line, in inverse proportion to the number of men, and in direct proportion to the other quantities, since we see that "the more" men "the less" time (hence *inverse*), but "the greater" thickness "the greater" time (direct), the higher, the greater time (direct), and the longer, the greater time (direct).

We first determine how many times greater or smaller the time would be, taking only one condition, such as the number of men, as different and assuming for the time that the other conditions, i. e., thickness, height and length, being equal.

We see that, all other things being equal, the increase in the number of men will make a proportional reduction in time, namely, in the above example, $\frac{6}{12}$ the length of time; again we see that if we consider only the thickness, the thinner wall will only take $\frac{2}{4}$ the length of time, hence both together will give $\frac{6}{12} \times \frac{2}{4}$ as much time as the given case. Again we see that the required wall is $\frac{9}{3}$ the height of the given wall, so that independently of the greater number of men and the thinner wall, the time will be $\frac{9}{3}$ greater on account of the greater height, giving $\frac{6}{12} \times \frac{2}{4} \times \frac{9}{3}$ the time; but finally the required wall is not so long as the given one by $\frac{60}{120}$.

The required time therefore is :

$$=\tfrac{6}{12}\times\tfrac{2}{4}\times\tfrac{9}{8}\times\tfrac{60}{120}\times7=2\tfrac{5}{8} \text{ days.}$$

It will be seen that the whole operation consists in the continued multiplication of a number of fractions (or ratios). The successive ratios are first multiplied together, and the product, which may be termed a compound ratio, is then multiplied by the quantity in which the change is required.

EXAMPLES :

(1) If 6 men can dig a ditch 40 yards long in 8 days, working 9 hours a day, how long a ditch can be dug by 4 men in 18 days, working 10 hours a day? All conditions of width, depth, soil, etc., are supposed to be the same in both cases.

Answer. $66\tfrac{2}{3}$ yards=66 yds., 2 ft.

Men	Days	Hours	Length
4	18	10	?
6	8	9	40 yds.

Here all quantities are in direct proportion, hence :

$$\tfrac{4}{6}\times\tfrac{18}{8}\times\tfrac{10}{9}\times40=66\tfrac{2}{3}.$$

(2) If 6 men working 9 hours a day, can build a wall 40 yards long in 8 days, how many days will be required for 4 men working 10 hours a day to build a wall $66\tfrac{2}{3}$ yards long?

Answer.

Men	Days	Hours	Yards	=3 Feet
4	?	10	$66\tfrac{2}{3}$	=200
6	8	9	40	=120

$$\tfrac{6}{4}\times8\times\tfrac{9}{10}\times\tfrac{200}{120}=18 \text{ days.}$$

The men are in *inverse* proportion, the more men the less time ; the hours are *inverse*, the more hours per day the fewer days ; the lengths are *direct*, the longer the wall the longer the time.

(3) 1500 men can subsist on the provisions in a fortress for 30 days if each man consumes 1 pound of bread per day. If now 500 additional men are thrown into the place and they are only required to hold out 24 days, what will be the allowance per man?

Answer. $\tfrac{1500}{2000}\times\tfrac{30}{24}\times1=\tfrac{15}{16}$ lbs.

(4) If 40 weavers can weave 200 pieces of goods, each piece 48 yards long and 40 inches wide, in 7 weeks, working 6 days a week and 12 hours a day ; how many pieces will be made by 60 weavers, working 8 weeks, 5 days a week and 8 hours a day, if each piece of goods be 36 yards long and 28 inches wide ?

Men	Pieces	Weeks	Days	Hours	Length	Width
60	?	8	5	8	36	28
40	200	7	6	12	48	40
Direct		Direct	Direct	Direct	Inverse	Inverse

Answer. $\frac{60}{40} \times 200 \times \frac{8}{7} \times \frac{5}{6} \times \frac{8}{12} \times \frac{48}{36} \times \frac{40}{28} = 362\frac{358}{441}$ pieces.

See Appendix § 322.

BOOK VIII.

Proportional Numbers (Ratios) and their Use.

65.

In many mathematical researches the actual values of the subjects are not required, but only their *relative* values, or the *ratio* which they bear to each other.

By *ratio* is meant, how many times greater or smaller one magnitude is than the other.

Instead of stating the actual values in numbers, we may divide one of them by the other, and thus use the quotient, or ratio, instead. For instance, if one person, A, has $200, and another one, B, has $600, we may say that the possession of the first person is to the second as 200 is to 600, or more briefly, as 2 is to 6; which may be written " as 2 : 6, the colon here being used as an abbreviation of the words "is to", as well as an indication of division. We will always have the same relation existing if we multiply or divide *both* numbers by the same quantity, and hence may have 1 : 3 as $\frac{1}{4} : \frac{3}{4}$, which is read : 1 is to 3 as $\frac{1}{4}$ is to $\frac{3}{4}$. The double colon (::) or the sign of equality (=) is usually used for the word "as" between two equal ratios, thus : 1 : 3 :: $\frac{1}{4} : \frac{3}{4}$, or 1 : 3 = $\frac{1}{4} : \frac{3}{4}$ and it is apparent that the first number divided by the second, gives the same result as the third divided by the fourth. This follows also if we regard the colon as a symbol of division (which it still is), for we have $\frac{1}{3} = \frac{\frac{1}{4}}{\frac{3}{4}}$, the truth of which is obvious, and hence we say that the two pairs of quantities are in the same *ratio*. This question of ratio is not limited to quantities; suppose there are four cities A, B, C, D, of which the populations are in the proportion (or ratio) of 3, 2, 6, and 4, or as written 3 : 2 : 6 : 4, the relation will be the same if we divide them all by any given number, and if we divide successively by 3, 2, 6, and 4 we shall have

$$
\begin{array}{ccccccc}
 & \text{A} & \text{B} & \text{C} & \text{D} \\
\text{as} & 3 & : & 2 & : & 6 & : & 4 \\
\text{or as} & 1 & : & \tfrac{2}{3} & : & 2 & : & \tfrac{4}{3} \\
\text{or} & \tfrac{3}{2} & : & 1 & : & 3 & : & 2 \\
\text{or} & \tfrac{1}{2} & : & \tfrac{1}{3} & : & 1 & : & \tfrac{2}{3} \\
\text{or} & \tfrac{3}{4} & : & \tfrac{1}{2} & : & \tfrac{3}{2} & : & 1
\end{array}
$$

That is, for example, the population of A is to C, as $3 : 6$ or as $1 : 2$, or as $\tfrac{3}{2} : 3$ etc., and the population of A is to D, as $3 : 4$, or $1 : \tfrac{4}{3}$ or as $\tfrac{1}{2} : \tfrac{2}{3}$, etc., etc.,

66.

We see also by the preceding section that when we have the ratios of several quantities expressed as numbers, that we know just how many times each one is greater or smaller than the other. Thus we see that B has $\tfrac{2}{3}$ times as many inhabitants as A, and that C has twice as many, and D $\tfrac{4}{3}$ as many. Also that A has $\tfrac{3}{2}$ as many inhabitants as B, while C has 3 times and D twice as many.

When therefore we have the ratios of a number of quantities given, and the actual value of one of them is known, we can easily find all the others. We have only, by inspection to select the ratio which belongs to the known quantity and divide all the others by it, thus making the ratio of the known quantity equal to unity. (See § 322)

Suppose again, for example, that we know the population of the four towns to be in the ratio of $3 : 2 : 6 : 4$, and know that A has 6000 inhabitants, we can find the population of the other towns as follows :

$$
\begin{array}{lcccc}
 & \text{A} & \text{B} & \text{C} & \text{D} \\
\text{are as} & 3 & 2 & 6 & 4 \\
\text{dividing by } 3 = & 1 & \tfrac{2}{3} & 2 & \tfrac{4}{3}
\end{array}
$$

whence since A=6000=6000 inhabitants

$$
\begin{aligned}
\text{B} &= \tfrac{2}{3} \times 6000 = 4000 \quad \text{``} \\
\text{C} &= 2 \times 6000 = 12000 \quad \text{``} \\
\text{D} &= \tfrac{4}{3} \times 6000 = 8000 \quad \text{``}
\end{aligned}
$$

EXAMPLES :

The age of a person, A, bears to that of a person, B, the ratio 2 : 5.   A is 20 years old, how old is B?

Answer. $\frac{5}{2} \times 20 = 50$ years.

67.

Ratios are often stated so that one of them is equal to 100, if as is often the case, it is desired to express them in percentages.   For example, the mark is to the franc as 25 is to 20, or as 100 : 80, so that the franc is worth 20 per cent. (symbol %) less than a mark, and any value in francs must be multiplied by $\frac{80}{100}$ or 0.80 to obtain the value in marks.

68.

Ratios may be used to divide a given magnitude into proportional parts.   This operation is best explained by a practical illustration.

Suppose it is required to divide the number 72 into 3 parts, which shall bear to each other the relation 2 : 4 : 6.   This is done by dividing the given number by the *sum* of the proportional numbers, and then multiplying the quotient by each of them separately.

Thus if we divide 72 by 2+4+6=12, we obtain the 12th part of 72.   If then we take this 12th, twice, 4 times, and 6 times, we shall have three products which still are in the given proportion, and also when added together make the given number. This gives also a proof of the correctness of the work, since $\frac{12}{12}$ should equal the whole amount.

The work in the example is as follows :

$$\text{The 1st part} = 2 \times \frac{72}{2+4+6} = 2 \times 6 = 12, \text{ i. e. } \tfrac{2}{12} \text{ of 72.}$$
$$\text{the second part} = 4 \times 6 = 24 \quad = \quad \tfrac{4}{12} \quad ''$$
$$\text{the third part} = 6 \times 6 = 36 \quad = \quad \tfrac{6}{12} \quad ''$$
$$\text{Proof total,} \quad 72 \quad = \quad \tfrac{12}{12} \quad ''$$

EXAMPLE : Four merchants enter into a speculation, each contributing the following amounts:

A, $600, B, $750, C, $1200, and D, $1050.   The venture resulted in a profit of $1500, which was divided in the same proportion as each had contributed.   How much did each receive?

Answer. A, $250, B, $312½, C, $500, and D, $437½.

Explanation : The total amount contributed was

$$600 + 750 + 1200 + 1050 = 3600$$

Hence we have $A = \frac{600}{3600} \times 1500 = \$250$

$$B = \frac{750}{3600} \times 1500 = \$312\tfrac{1}{2}$$
$$C = \frac{1200}{3600} \times 1500 = \$500$$
$$D = \frac{1050}{3600} \times 1500 = \$437\tfrac{1}{2}$$

The above example is based on the assumption that the money of all four men was in use the same length of time.

When the different amounts are not all invested in an undertaking for the same length of time, a *compound* proportion, including the various amounts of time, must be made.

For example : Suppose three persons enter into an undertaking ; A, contributes $6 for 2 months ; B, $12 for 4 months ; and C, $18 for 6 months. At the end of 6 months the undertaking is completed and results in a gain of $42, to be divided among the three men. The correct proportion is obtained in the following manner. A, has contributed $6 for 2 months, which is the same as $6 \times 2 = \$12$ for one month. This is clear, for it would have been just the same if one person had put in $6 for one month and then drawn out and a second person had put in $6 for the second month, and the contribution of two persons $6 for one month is the same as one person $12 for one month.

In the same way we see that the use of B's $12 for 4 months is the same as $12 \times 4 = \$48$ for 1 month ; and C's 18 for 6 months is $= 18 \times 6 = \$108$ for 1 month.

In this way we reduce the investments all to the same unit of time, i. e., one month, and can then proceed as before. Thus :

A, $6 for 2 months = $ 12 for 1 mo.
B, $12 " 4   "   = $ 48 "   "
C, $18 " 6   "   = $108 "   "

Total, $168

The total gain is $42, hence the shares are as follows :

$$A = 12 \times \tfrac{42}{168} = 12 \times \tfrac{1}{4} = \$3$$
$$B \qquad = 48 \times \tfrac{1}{4} = \$12$$
$$C \qquad = 108 \times \tfrac{1}{4} = \$27$$

Proof total, $42

EXAMPLE : In a business venture in which $200 was lost, the following interests were held : A put in $200 for 5 months ; B, $500 for 2 months ; C, $300 for 4 months ; and D, $600 for 3 months. In what proportion should the loss be shared?

Answer. A, $40 ; B, $40 ; C, $48 ; D, $72.

69.

Ratios, or Proportional numbers are of much use in determining the quantities to be used in various compounds and mixtures, the proportions being given, and always available for preparing any desired quantity of the mixture.

Such instances are constantly occurring in practical chemistry, technology, etc. A single example will make the method understood so that it can be applied to any case in actual practice.

EXAMPLE : Common gunpowder is a mixture of saltpetre, sulphur and charcoal, the proportions by weight being $16 : 2 : 3$, that is, 16 parts of saltpetre, by weight, must be mixed with 2 parts of the same weight of sulphur, and 3 parts of powdered charcoal. What will be the number of pounds of each material to make 1470 pounds of gunpowder?

Explanation : If we divide the total quantity, 1470, by $16+2+3=21$, we will obtain the number of times each proportional number must be taken to give the amounts of the corresponding materials. Hence 1470 lbs. of gunpowder will consist of

$$1470 \div 21 = 70$$
$$70 \times 16 = 1120 \text{ lbs. saltpetre}$$
$$70 \times 2 = 140 \text{ lbs. sulphur}$$
$$70 \times 3 = 210 \text{ lbs. charcoal}$$

Proof total, 1470 lbs. gunpowder

If we desire to state the percentage of each article in a given compound, we consider the total weight to be divided into 100 parts. By dividing 100 by the sum of the proportionate parts and multiplying the quotient by each of the proportionate numbers, we obtain the number of hundredths, i. e., the percentages. Thus in 100 parts of gunpowder we have :

$$\tfrac{100}{21} \times 16 = 76\tfrac{4}{21} \text{ parts saltpetre} = 76\tfrac{4}{21}\%$$
$$\tfrac{100}{21} \times 2 = 9\tfrac{11}{21} \text{ parts sulphur} = 9\tfrac{11}{21}\%$$
$$\tfrac{100}{21} \times 3 = 14\tfrac{6}{21} \text{ parts charcoal} = 14\tfrac{6}{21}\%$$

Total, 100 parts gunpowder

See Appendix § 323.

BOOK IX.

Weights and Measures.

70.

We have already assumed some familiarity on the part of the student with a few of the terms and values used in denominate numbers, such as pounds, miles, hours, minutes, seconds, etc.

The various denominate units differ in importance to different men, according to their kinds of work and their nationality, and it is rarely necessary for many tables to be memorized by any one person, each man becoming familiar by practice with those which he is most frequently called upon to use.

The principal tables of denominate quantities will here be given in condensed form, with a few words of explanation, and reference must be made to special works upon the subject for special information.

Measures of Length—United States and Great Britain.

12 inches=1 foot.
3 feet=1 yard=36 inches.
5½ yards=1 rod=16½ feet=198 inches.
40 rods=1 furlong=220 yards=660 feet.
8 furlongs=1 mile=320 rods=1760 yards=5280 feet.

Of the above, the inch and the foot are most frequently used by mechanics. The ordinary two-foot rule has the inches subdivided by the system of repeated halving, thus giving $\frac{1}{2}$, $\frac{1}{4}$, $\frac{1}{8}$ and $\frac{1}{16}$ of an inch; and this is sometimes carried as far as to include 32nds and 64ths. This system, however, is now being used principally by carpenters, builders, etc., while machinists are generally using scales, calipers and measuring tools which have the inch subdivided into 10ths, 100ths and 1000ths.

The yard is much used by shop keepers for measuring cloth, carpet and fabrics generally, and is by them also subdivided into halves, quarters and eighths.

For long distances the mile is universally used, and portions of a mile given either in furlongs and feet, or in halves and quarters.

For engineering measurements steel tapes are much used, 100 feet long, with the feet subdivided into 10ths, instead of inches, thus giving 10ths, 100ths and 1000ths of the length of the tape.

The mile given in the above table is called the *statute* mile, and is always used on land. The *nautical* mile, used only at sea, is equal to 6,080 feet, being about 15 per cent. longer than the statute mile.

The only other system of measures of length which is extensively used is the Metric System.

Metrical Measures of Length—Used generally on the Continent of Europe.

The unit is the Metre=39.37 inches.

The metre is subdivided decimally and multiplied decimally, as below :

1 millimetre=$\frac{1}{1000}$ metre=0.03937 ins.
1 centimetre=$\frac{1}{100}$ metre=0.3937 ins.
1 decimetre=$\frac{1}{10}$ metre=3.937 ins.
1 metre=39.37 ins.=3.2808 ft.
1 dekametre=10 metres=32.8087 ft.
1 hectometre=100 metres=328.0869 ft.
1 kilometre=1000 metres=3280.869 ft.=0.621 mile.

In using the metric system it is important to *think* of the metre as a main unit and the subdivisions as decimals of it. In mechanical and scientific work the metre and the millimetre are usually employed, and sometimes the centimetre, the decimetre more rarely. In the machine shop, for instance, measurements are usually given directly in millimetres, as 325 mm., not 3 dcm., 2 cm., 5 mm.

For longer distances the kilometre is used exclusively, and should be kept in mind as the unit of out-door measurement, with the metre, its $\frac{1}{1000}$ part, for all subdivisions ; the dekametre and hectometre being hardly used at all. It is very desirable that the student should learn the values of these measurements directly from the use of a metric scale, and *not* by transformation into English measures. When such transformations must be roughly made, however, it will be convenient to remember the following :

1 millimetre=$\frac{1}{25}$ inch, approximately.
1 decimetre=4 inches,   "
1 metre=3 ft., 3 ins., and $\frac{3}{8}$ ins., very closely.
1 kilometre=$\frac{5}{8}$ of a mile, nearly.

Measures of Weight—United States and British.

The commercial system is the *Avoirdupois*; the unit being the *pound* of 7000 grains.

The system for weighing gold and silver is called Troy Weight, of which the pound contains 5760 grains.

For medicines and drugs the Apothecaries system is used, the grain and pound being the same as in Troy Weight, but the subdivisions of the pound being different.

Avoirdupois or Commercial Weight.

1 dram=27.34375 grains
16 drams=1 ounce=437$\frac{1}{2}$ grains
16 ounces=1 pound=7000 grains.
28 pounds=1 quarter.
4 quarters=1 hundredweight=112 pounds.
20 hundredweight=1 ton=2240 pounds.

It will be noticed that the "hundredweight" (so called) is 12 pounds more than 100 pounds, this having been the allowance for loss in handling merchandise in old times. The Ton of 2240 pounds, is sometimes called the *long ton* in commerce as distinguished from the *short ton* of 2000 pounds. When no explanation is made, the long ton of 2240 pounds is the legal value of the ton, but in engineering calculations, such as the load upon a bridge, the pressure of a mass of earthwork, or the lifting capacity of a crane, it is customary to use the word ton to mean 2000 pounds. In practice a hundredweight (used as one word) means always 112 pounds, while a hundred pounds, means 100 pounds exactly.

Troy Weight.

1 pennyweight=24 grains.
20 pennyweights=1 ounce=480 grains.
12 ounces=1 pound Troy=5760 grains.

Apothecaries Weight.

1 scruple=20 grains
3 scruples=1 dram=60 grains.
8 drams=1 ounce=480 grains.
12 ounces=1 pound=5760 grains.

Measures of Weight—Metric System.

The Metric unit of weight is the gram, which is the weight of a cubic centimeter of pure water, and which is equal to 15.432 grains. The gram is subdivided and multiplied decimally as follows :

1 milligram=$\frac{1}{1000}$ gram=0.015432 grains.
1 centigram=$\frac{1}{100}$ gram=0.15432 grains.
1 decigram=$\frac{1}{10}$ gram=1.5432 grains.
1 gram=1 gram=15.432 grains.
1 dekagram=10 grams=154.32 grains.
1 hectogram=100 grams=1543.2 grains.
1 kilogram=1000 grams=2.2046 pounds.
1 myriagram=10000 grams=22.046 pounds.
1 metric ton=1000 kilograms=2204.6 pounds.

In practice many of these subdivisions and multiples are rarely used. The gram and the milligram, are used by chemists and physicists all over the world. The kilogram is used almost everywhere on the continent of Europe except in Russia, and its subdivisions are generally referred to as $\frac{1}{10}$ kilo, $\frac{1}{2}$ kilo, etc., instead of the tabular names, while the multiples are similarly named as 10 kilos, 100 kilos, etc. It will be noticed that the Metric Ton, or *Tonne* as it is written in France, is very nearly the same as the English long ton, so nearly that for ordinary commercial purposes they may be considered the same.

Measures of Volume

Measures of Volume *are not the same* in the United States and in Great Britain, and hence it should always be stated as to which is meant.

In the United States the systems for Liquid and for Dry Measures of volume are also different from each other, while in England both liquid and dry substances are measured by the same system.

Liquid Measure—U. S. A. Only.

The unit of volume is the *Gallon*=231 cubic inches. The gallon is subdivided and multiplied as follows :

4 gills=1 pint=28.875 cu. in.
2 pints=1 quart=57.750 cu. in.
4 quarts=1 gallon=231 cu. in.
63 gallons=1 hogshead.
2 hogsheads=1 pipe or butt.
2 pipes=1 tun.

Of the above measures the pint and quart are most frequently used. The *barrel* is not a standard volume, although in the U. S. and in England a wine barrel is supposed to contain $31\frac{1}{2}$ gallons, but in referring to a barrel in liquid measure the number of gallons it contains should be stated.

A cylinder 7 inches in diameter and 6 inches high contains almost precisely a gallon, and a gallon of pure water at its greatest density weighs 8.33888 pounds. Ordinarily it may be taken at 8.34 pounds. A cubic foot contains 7.48052 U. S. gallons.

Dry Measure—U. S. A. Only.

The unit of Dry Measure is the Bushel=2150.42 cubic inches. The bushel is subdivided as follows :

2 pints=1 quart=67.2 cu. ins.
4 quarts=1 gallon=268.8 cu. ins.
2 gallons=1 peck=537.6 cu. ins.
4 pecks=1 struck bushel=2150.42 cu. ins.

The *barrel* is not a legalized unit in dry measure, and its value should always be stated in gallons, or in pounds weight of the substance it contains. A barrel of flour is equal to 196 pounds.

British Measures of Volume.

In the British or Imperial system the same measures are used both for liquid and for dry measure. The unit of the system is the Imperial Gallon=277.274 cubic inches. This is intended to be equal to 10 pounds avoirdupois weight of pure water at a temperature of 62° Fah.

The Imperial Gallon is subdivided and multiplied as follows:

4 gills=1 pint=1.25 lbs. water.

2 pints=1 quart=2.50 lbs. water.

2 quarts=1 pottle=5.00 lbs. water.

2 pottles=1 gallon=10.00 lbs. water.

2 gallons=1 peck=20.00 lbs water.

4 pecks=1 bushel=80.00 lbs. water.

4 bushels=1 coomb=320.00 lbs. water.

2 coombs=1 quarter=640.00 lbs water.

The measures above the gallon are used for dry measures exclusively, and it is customary to state all quantities above the bushel in bushels.

Metric Measures of Volume.

The unit of volume is the Litre, which is equal to 1 cubic decimetre. This is subdivided and multiplied decimally as follows:

Liquid.

1 millilitre=$\frac{1}{1000}$ litre.

1 centilitre=$\frac{1}{100}$ litre.

1 decilitre=$\frac{1}{10}$ litre.

1 litre=1 litre.

1 decalitre=10 litres.

1 hectolitre=100 litres.

1 kilolitre=1000 litres.

The principal measure used is the litre itself, and in trade the $\frac{1}{2}$ litre is often used, this being a little more than a pint ($\frac{1}{2}$ litre= 1.056 pint), and so convenient that the fact of its not being a decimal equivalent is overlooked. For chemical and physical measurements the *cubic centimetre* is much used, and called by this name, c. c., and not *millilitre*, which latter it really is.

The unit of dry measure in the metric system is supposed to be the Stere=1 cubic metre, but in practice the term cubic metre is very generally used, and the subdivisions and multiples so named; i. e., $\frac{1}{10}$ cubic metre, 100 cubic metres, etc.

MONETARY SYSTEMS.

The various systems used for the money of different countries are too numerous to be described here, but a few of the most important will be given.

United States and Canada.

The unit is the dollar ($), subdivided and multiplied decimally. The dollar is divided into 100 cents, and the other units are as follows :

1 dime=10 cents=$\frac{1}{10}$ dollar.
1 dollar=100 cents.
10 dollars=1 eagle.

Besides these decimal units there are coins as follows :

$\frac{1}{4}$ dollar=25 cents.
$\frac{1}{2}$ dollar=50 cents.
Double eagle=20 dollars.

These coins are made for convenience, but are not known by their names in reckoning, the quarter and half-dollar being counted as 25 and 50 cents, and the double eagle, as well as the eagle, as so many dollars.

Great Britain.

The unit is the pound sterling, or sovereign ($\pounds$), subdivided as follows :

The penny=$\frac{1}{240}$ pound.
1 shilling=12 pence=about 24 cents.
1 pound=20 shillings=240 pence=about $4.86.

Besides these there are the following coins; half-penny=$\frac{1}{2}$ penny, crown=5 shillings, half-crown=2$\frac{1}{2}$ shillings, florin=2 shillings, but the calculations are all made in pounds, shillings and pence.

Latin Monetary Union.

On the Continent of Europe the following countries have formed themselves into the Latin Monetary Union, and use the same system, i. e. France, Belgium, Switzerland, Italy and Greece. The unit is the Franc, called Lira in Italy, and Drachma in Greece.

The franc is subdivided into 100 centimes, centesmi in Italy, lepta in Greece. There are also gold pieces of 20 francs, and silver coins =$\frac{1}{2}$ franc, besides minor coins of nickel, but these have no special names, all the reckoning being done in francs and 100ths. The equivalent value of the franc is about 19.3 cents.

Germany.

The unit is the Mark, equal about 24 cents, subdivided into 100 pfennigs. There are gold coins of 20 marks, but all the reckoning is done in marks and 100ths.

71.

The relations between units of measurement of various countries are given in extensive tables, to which reference must be made when such information is desired. By the use of such tables the various conversions are readily performed, as an example will show.

Suppose for example it is required to know the equivalent of 732 Russian sachines in metres, and that one sachine is equal to $2\frac{1}{3}$ English yards, and 1 metre equals 1.09363 English yards; these values being found in tables; we have :

$$732 \text{ sachines} = 2\frac{1}{3} \times 732 = \frac{7 \times 732}{3} \text{ yards.}$$

$$1 \text{ yard} = \frac{1}{1.09363} \text{ metres} = \frac{100000}{109363} \text{ metres.}$$

$$\text{hence } 732 \text{ sachines} = \frac{100000 \times 7 \times 732}{3 \times 109363} = 1561.772 \text{ metres.}$$

72.

The exchange value of the monetary units of different countries is constantly varying with the variations in trade and financial dealings, values being affected by the demand and supply. If we have to convert money of one country into that of another, we must find the correct rate of exchange from the bankers reports for the day, and it is often necessary to reckon through a chain of such rates to get the required value. An example will show how such a reduction can be made.

Example : What will be the value in dollars, of 125 rubles (St. Petersburgh) if 100 rubles=202 marks at Berlin, 100 marks at Berlin=59 gulden at Vienna, and 135 gulden at Vienna=54 dollars at New York?

Since we have :

100 rubles=202 marks.
100 marks=59 gulden.
135 gulden=54 dollars.

we have:

$$1 \text{ ruble} = \frac{202 \text{ M.}}{100}, \quad 1 \text{ mark} = \frac{59 \text{ guld.}}{100}$$

and $1 \text{ gulden} = \frac{\$54}{135}$, whence

$$125 \text{ rubles} = \frac{125 \times 202 \times 59 \times 54}{100 \times 100 \times 135} = \$59.59.$$

When such calculations are required to be made frequently they can be readily stated without liability to error in the following manner.

How many dollars=125 rubles
When 100 rubles=202 marks.
100 marks=59 gulden.
135 gulden=54 dollars.

Then divide the product of the right-hand numbers by the product of those on the left, for the answer; taking care always to shorten the work as much as possible by cancelling out all common factors.

73.

Besides the tables and terms already described, there are many other calculations made in trade and commerce which cannot be given here but which must be learned by actual experience. There are many words such as net, gross, rebate, tare, tret, etc., etc., for the meanings of which the student must refer to the dictionary.

There are two ratios however, which are of sufficient interest to be described here. The "fineness," so-called, of gold or silver, is determined by the number of parts of pure gold or silver there are in 1000 parts of the alloy. The metal is of course pure only when it contains no alloy whatever, and is then $\frac{1000}{1000}$ fine. The standard alloy for gold for U. S. coinage is 900 parts of pure gold and 100 parts alloy, and hence is $\frac{900}{1000}$ fine.

Of this alloy the gold dollar contains 25.8 grains, the eagle 258 grains, and the double eagle 516 grains.

The standard "fineness" for silver is also $\frac{900}{1000}$, and the standard dollar contains 412.5 grains of this alloy.

PART II.

General Arithmetic.

The Use of Symbols.

ALGEBRA.

BOOK X.

Reckoning with Universal Symbols.

74.

Calculating by means of universal symbols (usually letters) involves a knowledge of the use of *opposing* magnitudes (positive and negative quantities), and hence this subject will first be studied in the following sections, §§ 74 to 79.

When several quantities are to be added together, and then from the sum several other quantities are to be subtracted, the operations may be readily indicated by writing the quantities one after the other, prefixing the sign $+$ to those which are to be added, and the sign $—$ to those which are to be subtracted.

Suppose, for example, the numbers 9 and 12 are to be added together, and from their sum the numbers 5, 3 and 2 are to be substracted ; these operations can be indicated by writing :

$$+9+12—5—3—2$$

and this expression does not mean that 2 is to be subtracted from 3, or 3 subtracted from 5, but that the sum of *all* the figures preceded by the *minus* sign, is to be subtracted from the sum of *all* the figures preceded by the *plus* sign.

This point being thoroughly understood and always borne in mind, we see that it is quite immaterial in *what order* the quantities are written, as the order can have no influence upon the final result. It is customary in writing such an expression, to begin with a *plus* (or positive) quantity, as it is then unnecessary to write any sign before it, for a quantity is always understood to be plus (or positive) unless it is indicated to be minus (or negative). When the first quantity is negative, the minus sign must never be omitted, but for simplicity the plus sign may be omitted, and in general any quantity without a sign prefixed is understood to be positive.

The following statements are therefore all identical in value, the result being $=11$, although the figures are written in different orders :

$$9+12-5-3-2$$
$$12+9-3-2-5$$
$$9-5+12-2-3$$
$$-5+9-3+12-2$$

75.

In every expression consisting of several quantities of different *signs*, the quantities which are preceded by the sign $+$ are called *positive* (or direct), and those which are prefixed by the sign $-$ are called *negative* (or inverse) quantities. The signs $+$ and $-$ are called *opposing* signs, each being the inverse of the other. Each sign belongs to the quantity *before* which it stands.

Addition of Opposing Quantities.

In determining the *sum* of several positive and negative numbers, the following points must be carefully observed :

(1) If all the quantities have *like* signs, i. e., are all positive or all negative, they are all added together and the same sign prefixed to the sum. Thus we have :

$$6+4+3=13$$
$$-6-4-3=-13$$

(2) If the quantities have *different* signs, the positive quantities are added together, and the negative quantities also added together; then the smaller of the two sums is taken from the greater, and to the remainder is prefixed the sign of the greater sum.

Such a remainder, regardless of the fact that it is positive or negative, is called the *algebraic sum* of the quantities.

If the sums of the positive and negative quantities are equal to each other, it is evident that the value of the expression $=0$, since equal and opposite magnitudes cancel each other.

Thus, for example:

$$+5-5=0$$
$$-6+6=0$$
$$+8-5=3$$
$$-8+5=-3$$
$$9-5+12-2-3=11$$
$$8-10-6+3=-5$$

Subtraction of Opposing Quantities.

77.

If we have to subtract the expression $4-10+7$ from 9, we write it

$$9-(4-10+7)$$

enclosing the expression which is composed of several quantities, in a parenthesis.

In order to perform the subtraction thus indicated, we remove the parenthesis and convert the subtraction into an addition, by changing the signs of all the *positive* quantities in the parenthesis into *negative* and all the negative into positive quantities.

In the above example we would therefore change the $+4$ into -4, the -10 into $+10$, and the $+7$ into -7, and thus have the addition; example:

$$9-4+10-7=19-11=8$$

The proof of the correctness of a subtraction is the fact that the remainder added to the subtrahend must equal the minuend. Thus, for example:

$$9-(+4-10)=9-4+10$$

must be correct because if we add the subtrahend $(+4-10)$ to the remainder $9-4+10$, we get:

$$9-4+10+4-10=9$$

which is equal to the minuend.

Again suppose it is required to subtract $-13+25-6$ from $-7+2$, we state the problem thus:

$$-7+2-(-13+25-6)=$$
$$-7+2+13-25+6=21-32=11$$

Multiplication of Opposing Quantities.

78.

If it is required to multiply $+7$ by $+3$ it is evidently $= 7 \times 3 = 21 = +21$. Now if we have to multiply -7 by $+3$, it is the same as taking -7 three times, or $-7-7-7 = -21$.

But what will it be if we are required to multiply -7 by -3?

If $+7$ times $-3 = -21$, then -7 times -3 must be equal to its exact opposite, i. e. $= +21$.

Hence we have the following rule: *Like signs give plus, unlike signs give minus.*

EXAMPLES: *

$$-5 \cdot -2 = +10$$
$$9 \cdot -1 = +9 \cdot -1 = -9$$
$$-2 \cdot 5 = -2 \cdot +5 = -10$$
$$3 \cdot 4 = +3 \cdot +4 = +12 = 12$$
$$-6 \cdot -5 \cdot -4 = +30 \cdot -4 = -120$$

Division of Opposing Quantities.

79.

The same rule holds good for division as for multiplication: *Like signs give plus, unlike signs give minus.* Thus:

$$\frac{-12}{-4} = +3, \text{ because } +3 \cdot -4 = -12, \quad (\text{See § 11.})$$

EXAMPLES:

$$\frac{3}{-2} = \frac{+3}{-2} = -\tfrac{3}{2} = -1\tfrac{1}{2}$$

$$\frac{-8}{6} = \frac{-8}{+6} = -\tfrac{8}{6} = -1\tfrac{1}{3}$$

$$\frac{+2}{+10} = +\tfrac{2}{10} = \tfrac{1}{5}$$

$$\frac{5 \cdot -7}{-2} = \frac{-35}{-2} = +17\tfrac{1}{2}$$

Those who wish to go more fully into the subject of positive and negative values are referred to § 320 (appendix.)

---

*Hereafter we shall use the *point* ($\cdot$) to indicate *multiplication*, and it need not be confused with the decimal point, since it will be placed *above* the line, not *on* the line.

The use of Letters as Symbols.

80.

Complicated calculations may often be very much simplified if the operations of addition, subtraction, multiplication etc., are not *performed* at once, but first merely *indicated* by the signs $+$, $-$, etc., so that we have the entire operation before us as a whole, before any of the detailed work is done, and can thus decide the best method of shortening the work by cancelling, collecting like members together, factoring etc., etc. As an example of the advantages of this method of proceeding, the following simple instance will serve.

Question? What will be the result, if we add together the half-sum and the half-difference of the fractions $\frac{4578}{8777}$ and $\frac{213}{7691}$?

Solution. We first proceed to indicate the operations stated by the question. We have then $\frac{4578}{8777} + \frac{213}{7691}$ as the *sum*, and $\frac{4578}{8777} - \frac{213}{7691}$ the *difference* of the two fractions, and hence the half-sum and half-difference will be represented respectively by

$$\tfrac{1}{2}(\tfrac{4578}{8777} + \tfrac{213}{7691}) \text{ and } \tfrac{1}{2}(\tfrac{4578}{8777} - \tfrac{213}{7691})$$

Both these quantities added together, is indicated by

$$\tfrac{1}{2}(\tfrac{4578}{8777} + \tfrac{213}{7691}) + \tfrac{1}{2}(\tfrac{4578}{8777} - \tfrac{213}{7691})$$

which then is the full *statement* of the question, in such a form that we can study it as a *whole*.

The first thing to do is to see how we can collect the various quantities together so as to simplify the somewhat complicated looking expression, and for this purpose we remove the parenthesis. This we can readily do, since the $\frac{1}{2}$ of each of the quantities within the parenthesis is the same as the $\frac{1}{2}$ of them taken together, and we have

$$\tfrac{1}{2} \cdot \tfrac{4578}{8777} + \tfrac{1}{2} \cdot \tfrac{213}{7691} + \tfrac{1}{2} \cdot \tfrac{4578}{8777} - \tfrac{1}{2} \cdot \tfrac{213}{7691}$$

We can now see at once that of the four members in the above expression there are two alike of the *same* sign and two alike of opposite signs. The latter, i. e. the 2nd and 4th members being equal in value and of opposite signs cancel each other, and may be struck out. The 1st and 3rd members, we see are the same and since each of them is equal to $\frac{1}{2}$ of the fraction $\frac{4578}{8777}$ the two together will be equal to the whole of that fraction, so that the answer to the question is $\frac{4578}{8777}$, obtained entirely by this reduction and without any reckoning whatever.

Now in examining such a problem as that just given, we see that in the various operations which have been performed, no changes whatever have been made in the original quantities themselves. The fractions above given have been shifted in position, multiplied, divided, added and subtracted, but to all intents and purposes we might, instead of writing out the quantities themselves, have used some simple and distinctive *symbol* for each original quantity, and then when the work was completed, given each symbol its original value again.*

Such symbols can have the operations indicated by the signs $+$, $—$, etc., and relieve the detail of the work entirely from the burden of bulky numerical quantities, and thus constitute a sort of mathematical short-hand, with much economy of labor and fewer opportunities for error.

The nature of the symbols to be employed is entirely a matter of choice. Ordinarily, however, it is customary to use the letters of the alphabet, since these are already well known and very convenient for the purpose.

The use of such symbols can be well explained by applying them to the same example given above. Suppose then again that we have to find the result of the addition of the half sum and half difference of the fractions $\frac{4579}{8774}$ and $\frac{213}{7691}$.

We may use the letter a, for the first fraction, and b, for the second fraction. In this case the sum of the two quantities will be $a+b$, and their difference $a—b$, and the half sum $=\frac{1}{2}(a+b)=\frac{a+b}{2}$ and the half-difference $=\frac{1}{2}(a—b)=\frac{a—b}{2}$ and hence the sum of the two will be $\frac{a+b}{2}+\frac{a—b}{2}$ which shows us the whole operation in one statement. Before substituting the values of the letters again we will see what reductions can be made. We can (according to § 20) unite the above

$$\frac{a}{2}+\frac{b}{2}+\frac{a}{2}-\frac{b}{2}$$

We then see at once that $+\frac{b}{2}$ and $-\frac{b}{2}$ cancel each other,

leaving $\frac{a}{2}+\frac{a}{2}$ which is clearly equal to a. But a is the symbol

---

*A practical example of such symbols is found in the use of bank checks, for instance, which may pass through several hands before being presented for payment, or in a less reputable manner perhaps, by poker chips, which are symbols of value passing through all the vicissitudes of the game to be redeemed at last.

which we selected for the fraction $\frac{4544}{8777}$, and hence this is the answer to the problem, as before.

81.

Second Example. A pedestrian walks at a uniform rate from a place, O, to another place, W. He starts 4 minutes, 59 seconds after eight o'clock in the morning from O, and arrives at W at 57 minutes, 48 seconds after 3 o'clock in the afternoon. What time of day is it when he is just half-way between the two places?

Solution. If we take $\frac{1}{2}$ the total amount of time he spends between the two points and add it to the time at which he started, we shall obtain the required hour at which the journey will be half completed. But in order to obtain the total time spent on the whole journey we must subtract the hour of starting from 12 to find out how much time elapsed before noon, and then add this to the time spent after noon; this sum is then divided by 2, and the result added to the hour of starting. The work involved is shown below :

O————————— : —————————W

8 h., 4 min., 59 sec.   3 h., 57 min., 48 sec.

	12^h	$0'$	$0''$
Starting time, A, M.,	8	4	59
Elapsed time to noon,	3^h	$55'$	$1''$
Elapsed time after noon,	3	57	48
Total time of journey,	7^h	$52'$	$49''$
Half the total time,	3^h	$56'$	$24\frac{1}{2}''$
Add the starting time,	8	4	59
Answer,	12^h	$1'$	$23\frac{1}{2}''$

Hence the traveler was just half-way between the two points at 1 minute, $23\frac{1}{2}$ seconds after 12 o'clock.

Now let us see how much this computation can be shortened by the use of symbols.

If we call the hour of starting $=a$, then the number of hours elapsed before noon will be $12-a$. Then if we call the hour the journey ends $=b$, we have for the whole time $12-a+b$, and also for the half time $\dfrac{12-a+b}{2}$.

If we add to this the symbol for the hour of starting, we have for the statement of the whole question : .

$$\frac{12-a+b}{2}+a \qquad (1)$$

This expression can doubtless be simplified by bringing it to a common denominator. Instead of leaving the second member without a denominator we may write it $\frac{2a}{2}$, and thus we get:

$$\frac{12-a+b}{2}+\frac{2a}{2} \qquad (2)$$

or, what is the same thing :

$$\frac{12-a+b+2a}{2} \qquad (3)$$

We see in the numerator two opposed quantities, namely $-a$ and $+2a$. But $-a+2a=-a+a+a=a$. We may therefore put for $-a+2a$ the simpler value a, which gives :

$$\frac{12+a+b}{2} \qquad (4)$$

All four of the above expressions must give the same result when we substitute for the letters their respective values. The last one, however, being the simplest, will give the least numerical work, as it can be reduced no further.

On inspection we see that it means that we must add 12 to the sum of the time of starting and the time of arriving and divide the result by 2. The actual work is as follows :

	12^h	$0'$	$0''$
$12=$	12^h	$0'$	$0''$
$a=$	8	4	59
$b=$	3	57	48
$2\,)$	24	2	47
	12^h	$1'$	$23\frac{1}{2}''$

It is hoped that a careful study of these examples and the explanations which have been given of them will at least give the student such an idea of the use of letters as symbols that he he can clearly understand the following section, and pursue his studies without meeting the difficulties and obscurity which so often embarrass and hinder the beginner.

General Ideas about the Use of Letters in Calculation.

82.

(1) When we have to make calculations of one quantity with another, it is most important first to consider how we may do the work with the least amount of actual reckoning; that is, with the least numerical and laborious work. In order therefore to avoid all unnecessary numerical work, such as addition, multiplication, etc., etc., and reduce the work as much as possible, as well as to abridge the labor of frequent writing of numbers, we may use *letters* as *symbols* instead of the actual quantities themselves.

(2) The result obtained by solving a problem when using symbols, applies to all problems of the same kind, independently of the numerical values of the original data. Thus, for instance, we found in the second example in the preceeding section that the desired result, expressed in symbols, was $\frac{12+a+b}{2}$.

Now this expression holds good for any example of the same sort, independently of the actual hours of starting and arriving given in the above example, and if another hour of starting be substituted for a, and a different time of arrival for b, the result will still give the time of the middle of the journey. Hence we see that the expression $\frac{12+a+b}{2}$ is a *general* rule, or *formula*, in which, if we substitute any given values, we may obtain easily and correctly the required result. Suppose for instance, the traveler should start from O at 9 o'clock A. M., and arrive at W at 3 o'clock P. M., we can at once find the time when he is at the middle of the journey, simply, by making $a=9$ $b=3$, and we have $\frac{12+9+3}{2}=12$ noon.

We thus see that the letters do *not* have *absolute* values but sometimes one value and sometimes another, as we may choose.

(3) There is great economy both of time in writing and in speaking, by using symbols instead of numbers, thus obtaining, as has been already said, a kind of mathematical shorthand.

Besides these advantages there is the very important feature of being able to see the whole operation at once before the eyes, and follow the transformation and action of the various quantities; which feature is almost entirely lost with arithmetical operations. This ability to examine a problem as a whole also enables one to see how many important reductions and simplifications may be made.

When any problem, which has been stated in a general form with symbols, is solved, the solution applies to all problems of the same kind merely by inserting the data for each special case.

Such a *general* solution differs from a *special* result, because the magnitudes of the various data are left undetermined, and by the various operations possible with symbols the expression may be reduced to its simplest form directly.

A general solution also teaches much more about the problem in hand, as it gives a broad general idea of the subject, showing causes and effects, before the eyes in a manner not possible with general solutions.

83.

If these fundamental ideas are clearly understood, the student should find this chapter upon the use of letters as symbols one of the easiest in the whole subject of mathematics, since all reductions are made in the most mechanical and least laborious manner possible.

The operations with letters are not *numerical* calculations at all, but rather a kind of reckoning in which the various operations are rather *indicated* than performed, the symbols being placed, collected, separated and generally manipulated quite independently of the actual values which may be assigned to them. The definite values are not given to the symbols until the calculations are completed, after all possible reductions have been made and the expressions reduced to the very simplest form.

The manner of making such reductions and performing the various operations will be fully explained in the following sections.

84.

In order to make the operations more agreeable to the eye, it is preferable to place the letters in the order in which they occur in the alphabet. For instance, the addition of the quantities c, a and b is written $a+b+c$, rather than $c+a+b$.

85.

To indicate that two letters are to be multiplied together, the two factors are written closely together without using any symbol of multiplication whatever, these symbols being omitted as unnecessary. Thus, if b, a and x are to be multiplied together, instead of writing $a \cdot b \cdot x$, we write simply abx. In the same manner 3 multiplied by a and by y is written $3ay$. When a letter is to be multiplied by a number, the *numerical* factor, which is called the *co-efficient*, is always written *first*, Thus a multiplied by 3 is written $3a$, and 3 is the coefficient of a. In $\frac{2}{3}ax$, $\frac{2}{3}$ is the coefficient of ax. In $\frac{4ab}{5}$ (that is $\frac{4}{5}ab$), $\frac{4}{5}$ is the coefficient of ab.

The coefficient 1 is never written, but simply understood, as $1 \cdot 8 = 8$; in like manner we write for $1 \cdot a$, or $1 \cdot abc$, simply a and abc. In $\frac{a}{5b}$ the coefficient is really $\frac{1}{5}$, for $\frac{1 \cdot a}{5 \cdot b} = \frac{1}{5} \cdot \frac{a}{b}$.

86.

Expressions in letters are said to be *similar* when they consist of the same letters, arranged in the same manner, although the coefficients may be different. Thus, for example, $2ab$, $-\frac{2}{3}ab$ and $12ab$ are similar quantities. So likewise are $3aax$ and $25aax$. The expressions ab, $a+b$, $3abc$ and $2abb$ are dissimilar, being composed of different letters or in different arrangements.

Addition.

87.

We know that $7 + 7 + 7 = 3 \cdot 7$, and in like manner we see that $a + a + a = 3a$. We also see that $3a + 2a = a + a + a + a + a = 5a$, and hence for addition we have the rule : Write the symbols which are to be added together in any order which may be convenient, prefixing to each symbol its proper sign, and collecting all *similar* expressions by taking the *algebraic sum* of their coefficients.

Thus, $7x - 10x = -3x$, since the algebraic sum of 7 and -10 $= -3$;

$$abc + 6abc = 1 \cdot abc + 6abc = 7abc \; ;$$
$$\tfrac{2}{3}ax + \tfrac{3}{4}ax = 1\tfrac{5}{12}ax \; ;$$
$$2ab + 3ax + ab = 3ab + 3ax \; ;$$
$$a - 5a = -4a.$$

When the quantities to be added contain various members, it is better to arrange them so that similar expressions are over each other, as a general view of the whole operation can then be obtained. Thus, for example, to add together the following : $8ax+3bc$, $2aax-3$, $-5x+7$, and $6bc-4ax$, we have :

$$8ax+3bc+2aax-3-5ax+7+6bc-4ax=2aax-ax+3bc+4$$

but this would be much clearer arranged as follows :

$$\begin{array}{l} 8ax-3bc \\ \qquad +2aax-3 \\ -5ax \qquad\quad +7 \\ -4ax+6bc \\ \hline \end{array}$$

Total, $\quad -ax+3bc+2aax+4$

in which we have placed all the ax's over each other, the bc's over each other, and the other quantities also each in columns of their own.

Subtraction.

88.

In subtraction we proceed as in § 77, *changing the signs* of the members of the subtrahend and then taking the *algebraic* sum of minuend and subtrahend, also collecting all simular expressions. For example, to subtract $2a$ from $6a$, we have $6a-2a=4a$. Again $2ax$ from $6ax$ gives $6ax-2ax=4ax$; or b from a gives $a-b$.

If it is desired to *indicate* that any quantity composed of several members, is to be subtracted from any other quantity, the subtrahend is enclosed in a parenthesis and a minus sign prefixed; and if the parenthesis marks are removed, the signs of every member in it must be reversed.

For example, to subtract $y-a$ from x, we *indicate* the operation thus :

$$x-(y-a)=x-y+a$$

the signs of y and a being changed when the parenthesis is removed.

Again to subtract $2ab+6bc-4x$ from $3ab-3bc$, the operation is stated thus:

$$2ab-3bc-(2ab+6bc-4x)$$

In like manner

$$a+b-(a-b)=a+b-a+b=2b$$

If the minuend and subtrahend both consist of several members, we may write the subtrahend under the minuend, like members under like, and then convert the problem into an example in addition by changing the signs of the members of the subtrahend. This will be understood by reference to § 77, where it will be seen that to subtract $+4$ and -10 is the same as to add -4 and $+10$.

$$
\begin{array}{ll}
\text{Thus, from} & 2ab-3bc \\
\text{subtract} & 2ab+6bc-4x \\
& \quad (-) \quad (-) \ (+) \\
\hline
\text{difference} & \quad\quad -9bc+4x
\end{array}
$$

$$
\begin{array}{ll}
\text{again, from} & 2ax-6bz+7 \\
\text{subtract} & 6ax-3by-3 \\
\hline
& -4ax-6bz+3by+10
\end{array}
$$

The correctness of the subtraction is proved if the difference added to the subtrahend gives the minuend again.

Multiplication

89.

First Case. If the factors each consist of but one member, the letters are written next to each other in alphabetical order, and preceded by the product of the numerical coefficients. § 320.

For example $2a$ multiplied by $3b$, gives $2a \cdot 3b = 2 \cdot 3 \cdot ab = 6ab$. In the same way we have :

$$3ab . 5ac = 15aabc$$

$$\tfrac{2}{3}ab \cdot \tfrac{3}{4}x = \tfrac{1}{2}abx$$

$$\tfrac{1}{3}a \cdot \tfrac{3}{4}bc = \frac{abc}{4}$$

$$\tfrac{2}{3}ax \cdot \tfrac{3}{4}b \cdot \tfrac{4}{8} = \frac{abx}{3}$$

$$17ax \cdot 3b = 51abx$$

$$3abx \cdot 3bx = 9abbxx$$

$$5a \cdot \tfrac{1}{8}b \cdot \tfrac{4}{3}c = \frac{2abc}{3}$$

$$\tfrac{2}{4}am \cdot \tfrac{7}{8}n = \frac{amn}{4}$$

Second Case. If one factor contains one member and the other factor several members, the multiplication is merely *indicated* by enclosing the factor which consists of several members* in a parenthesis and writing the other factor outside of the parenthesis. If in the course of reduction the parenthesis is to be removed, this is done by multiplying each member in the parenthesis by the single factor. Thus if the quantity $a+b-c$ is to be multiplied by a, we indicate the operation by writing $a(a+b-c)$ and performing the multiplication we have :

$$a(a+b-c)=aa+ab-ac.$$

In like manner we have :

$$2a(b-3c)=2ab-6ac\ ;$$
$$3ab(7ab-3ax+1)=21aabb-9aabx+3ab\ ;$$
$$(9ax-\tfrac{5}{8}by)7b=63abx-5bby\ ;$$
$$4+5(x+6)=4+5x+5\cdot6=34+5x\ ;$$
$$x(y-1)=xy-x\ ;\ \ (a+1)b=ab+b\ ;$$
$$2ac-ab+a(b-c+x)=2ac-ab+ab-ac+ax=ac+ax$$

If the single factor has a minus sign, the rule of §78 is always to be applied. Thus :

$$a+7-3(a-2b+4)=?$$

Here we multiply —3 and $+a=$ —$3a$
$$-3 \text{ and} -2b= +6b$$
$$-3 \text{ and} +4= 12$$

hence we have

$$a+7-3a+6b-12=6b-2a-5$$

Third Case. If both factors are polynomials (i. e., consist of more than one member), we multiply *each part* of one factor by *each part* of the other factor, and then collect all the similar members together. It will usually be found convenient to select for the multiplier the factor which is composed of the fewer parts. In order to follow the work clearly, and avoid errors it is advisable to write the factors under each other, and sum up the products as shown in the following examples, 1, 2, and 3.

It is quite immaterial with which part of the multiplier the first multiplication is performed. Suppose we have $a+b-c$ to be

---

*A quantity composed of more than one member, the members being connected by the sign + or —, is called a *polynominal*.

multiplied by $a-b$, we have, multiplying each part of $a+b-c$ first by a and then by $-b$:

$$(a-b)(a+b-c)=aa+ab-ac-ab-bb+bc$$
$$=aa-ac-bb+bc$$

or as follows:

(1) $\quad a+b-c$
$\quad\quad a-b$
_____
$\quad aa+ab-ac$
$\quad\quad -ab-bb+bc$
_____
$\quad aa-bb-ac+bc$

(2) $\quad a+b$
$\quad\quad a-b$
_____
$\quad aa+ab$
$\quad\quad -ab-bb$
_____
$\quad\quad aa-bb$

(3) $\quad a-b$
$\quad\quad a+b$
_____
$\quad aa-ab$
$\quad\quad +ab-bb$
_____
$\quad\quad aa-bb$

In the same way we perform the following multiplications:

$$(a+b)(a+b)=aa+ab+ab+bb$$
$$=aa+2ab+bb;$$

$$(a-b)(a-b)=aa-ab-ab+bb$$
$$=aa-2ab+bb;$$

$$(x+1)(y-1)=xy-x+y-1;$$

$$(3ax-4by)(5bx+1)=15abxx+3ax-20bbxy-4by;$$

$$(\tfrac{2}{3}x-\tfrac{1}{4}y)(3x-\tfrac{4}{3}y-\tfrac{3}{2})=2xx-\tfrac{121}{18}xy-x+\tfrac{11}{4}yy+\tfrac{15}{8}y.$$

Factoring.

90.

Besides having to multiply several factors together, we often have the reverse problem to solve; namely, that of separating a given expression into its factors, and as factoring is very often the most valuable and important method of reduction, the student must become thoroughly familiar with it and give the subject his most earnest attention. The following examples will show how factoring is performed.

Suppose we have multiplied $3a-4c+2$ by $3ab$, we obtain the product $9aab-12abc+6ab$.

If, now, we had the reverse problem given, we could see by careful inspection, first, that every member of this last expression could be divided by the numerical factor 3, without a remainder, and also would observe that the literal factor ab is found in every member. We can therefore take out the common factor $3ab$ from each member and write it in front of a parenthesis in which is enclosed the balance of the expression, thus :

$$9aab-12abc+6ab=3ab(3a-4c+2)$$

In the same manner we can see by inspection of the expression $xx+xy$, that it is composed of two factors, one of which is x, since x appears as a multiplier in both members, and that :

$$xx+xy=x(x+y)$$

There is one form of expression which sometimes causes difficulty to beginners, namely, that of the form $yx+y$. The common factor here is at once seen however, if we write it $yx+y\cdot1$, so that

$$yx+y=yx+y\cdot1=y(x+1)$$

In the same way we see that

$$x-xx=x(1+x);$$
$$\text{and}\quad x+mx+nx=(1+m+n)x$$

The correctness of any problem in factoring can be proved by multiplying the factors together again ; the result should be the original expression.

Further examples are :

$$3ab+3ax=3a(b+x);$$
$$2Rn-4R=2R=(n-2);$$
$$y-yz=y(1-z);$$
$$\frac{ah}{2}+\frac{bh}{2}=\frac{h}{2}(a+b)=h\frac{a+b}{2}$$

If the first member of the expression to be factored is preceded by a minus sign, the expression should be enclosed in a parenthesis, preceded by a minus sign, before taking out the common factor. Thus for example :

$$a—bc—bd+be=a—(bc+bd—be)$$
$$=a—b(c+d—e)$$

The reason for changing the signs of the quantities in the parenthesis will be clear when it is seen that they will all be changed back if the multiplication by $—b$ is performed. Again :

$$5ax—3by+6bz=5ax—3b(y—2z)$$
$$=5ax+3b(2z—y)$$

Here we see that $—3b(y—2z)$ is the same as $+3b(2z—y)$, since the same result will be given in both cases when the multiplication is performed.

A complicated expression may often be greatly simplified by means of factoring. Thus, in the following expression, the first two members have the common factor ac and c, and the two following have the factor bc. Taking these factors out the following reduction can be made :

$$abc—acc—bbc+bcc+abd—acd—bbd+bcd$$
$$=ac(b—c)—bc(b—c)+ad(b—c)—bd(b—c)$$

Now we see that each member contains the binomial* factor $(b—c)$. By taking out this common factor we have :

$$=(b—c)\ [ac—bc+ad—bd]$$

We now see that the quantity in the second parenthesis can be factored, since the first two members both contain c, and the last two contain d, hence

$$=(b—c)[c(a—b)+d(a—b)]$$

and factoring it again, we get

$$=(b—c)[(a—b)(c+d)]$$
$$=(a—b)(b—c)(c+d)$$

which is a great simplification of the original expression.

91.

When a quantity is multiplied by itself we call the product the *square* of the quantity. The reason for this name will be seen when the student takes up geometry.

---

* " *Binomial* " means consisting of *two* quantities connected by the signs + or —, just as *polynomial* means consisting of several such.

For Example : $8+8=64$, which is the *square* of 8; the square of 5 is 25; the squares of 1, 2, $\frac{1}{2}$, $\frac{1}{3}$, $\frac{2}{3}$ are 1, 4, $\frac{1}{4}$, $\frac{1}{9}$, $\frac{4}{9}$ respectively. The square of $\frac{a}{2}$ is $\frac{aa}{4}$ (which is read "$\frac{1}{4}$ of a square," or "a square divided by 4.") Remembering this definition we have the two following important rules to learn by heart:

I. *The sum of two quantities* $(a+b)$ *multiplied by their difference* $(a-b)$ *is equal to the difference of their squares.*

Thus if we perform the multiplication, we have :

$$(a+b)(a-b)=aa+ab-ab-bb=aa-bb$$

By memorizing this rule, we are able at once to write down the product of the sum and difference of two quantities without performing the intermediate work.

EXAMPLES :

$$(a+x)(a-x)=aa-xx$$
$$(1+x)(1-x)=1-xx$$
$$(a+1)(a-1)=aa-1$$
$$(2a+\tfrac{1}{2}b)(2a-\tfrac{1}{2}b)=4aa-\frac{bb}{4}$$

(2) *When a quantity consists of the difference of two squares,* (as $aa-bb$; $bb-yy$, &c.,) *it can always be separated into two factors, consisting of the sum and difference of the quantities.*

This rule is simply the reverse of the preceding one.

EXAMPLES :

$$xx-zz=(x+z)(x-z)$$
$$xx-aa=(x+a)(x-a)$$
$$zz-1=(z+1)(z-1)$$
$$1-xx=(1+x)(1-x)$$
$$4aa-\frac{bb}{9}=\left(2a+\frac{b}{3}\right)\left(2a-\frac{b}{3}\right)$$
$$57\times57-43\times43=(57+43)(57-43)=100\times14$$

Division.

92.

Division is always *indicated* by writing the quantities in the form of a fraction. For example : If a is to be divided by b, the fact is indicated by writing $\frac{a}{b}$ (read "a divided by b," or more briefly "a over b ").

In the same manner $2ax$ divided by $3by$ is written $\dfrac{2ax}{3by}$, and $a+b$ divided by $a-b$ is written $\dfrac{a+b}{a-b}$.

(1) If the divisor and dividend have any common factors, these factors may at once be stricken out, or cancelled, since they do not affect the quotient. (See §§ 22, 23, 38.) Thus:

$$\frac{ac}{c}=d\,; \quad \frac{2abc}{3bc}=\frac{2a}{3}\,; \quad \frac{2abx}{3aabx}=\frac{2}{3a}\,; \quad \frac{6abb}{4bc}=\frac{3ab}{2c}$$

$$\frac{xy}{6y}=\frac{x}{6}\,; \quad \frac{a}{a}=1\,; \quad \frac{b}{b}=1\,; \quad \frac{a+b}{a+b}=1\,; \quad \frac{a+c-d}{a+c-d}=1\,;$$

$$\frac{(a+b)\,(a-b)}{a-b}=a+b\,; \quad \frac{ab\,(a+b)}{2b\,(a+b)}=\frac{a}{2}\,; \quad \frac{m\,(a+b)}{m\,(a+c)}=\frac{a+b}{a+c}$$

(2) If the dividend is composed of several members (polynomial) and the divisor consists of but one member (monomial), the divisor can be divided into each member of the dividend. (See § 20.)

Thus we have for example:

$$\frac{a+b-c}{d}=\frac{a}{d}+\frac{b}{d}-\frac{c}{d}\,;$$

$$\frac{8abb-10ax}{2ab}=\frac{8abb}{2ab}-\frac{10ax}{2ab}=4b-\frac{5x}{b}$$

(3) If both dividend and divisor are polynomials the division may be performed in two ways: (a) By factoring both quantities, whenever this is possible. This usually renders the division much simpler.

EXAMPLES:

$$\frac{ab-ax}{bb-bx}=\frac{a(b-x)}{b(b-x)}=\frac{a}{b}\,;$$

$$\frac{ax+xx}{3bx-xx}=\frac{x(a+x)}{x\,(3b-x)}=\frac{a+x}{3b+x}\,;$$

$$\frac{y-yy}{3by-3byy}=\frac{y(1-y)}{3by\,(1-y)}=\frac{1}{3b}\,;$$

$$\frac{ac-bcx-cz}{9bcz-cz}=\frac{c(a-bx-z)}{cz(9b+1)}=\frac{a-bx-z}{z(9b-1)}\,;$$

$$\frac{6aa-3ab}{12ac-6bc}=\frac{3a(2a-b)}{6c(2a-b)}=\frac{a}{2c};$$

$$\frac{4ax-3bx-xx}{8ayz-6byz-2xyz}=\frac{x(4a-3b-x)}{2yz(4a-3b-x)}=\frac{x}{2yz};$$

$$\frac{ay+ty-av+tv}{az+tz-2au+2tu}=\frac{y(a+t)-v(a+t)}{z(a+t)-2u(a+t)}=$$
$$=\frac{(y-v)(a+t)}{(z-2u)(a+t)}=\frac{y-v}{z-2u};$$

$$\frac{(a+b)(a+b)}{aa-bb}=\frac{(a+b)(a+b)}{(a+b)(a-b)}=\frac{a+b}{a-b}\ (\text{see §§ 91, 92});$$

$$\frac{xx+x}{1-xx}=\frac{x(x+1)}{(1+x)(1-x)}=\frac{x}{1-x};$$

$$\frac{mz-m}{m-my}=\frac{m(z-1)}{m(1-y)}=\frac{z-1}{1-y}=\frac{(-1)(z-1)}{(-1)(1-y)}=\frac{1-z}{y-1}.$$

(*b*) The second method is that of so-called "partial division," which may be deferred to the Appendix. (See § 321.)

Fractions.

When we have to work with fractional expressions, the general rules and methods for fractions are used, as explained in the following sections, §§ 93, 94. See Appendix, §321.

Addition and Subtraction of Fractions.

93

(1) If the fractions have a common denominator, we simply take the algebraic sum or difference of the numerators, and place the result over the common denominator. Thus for example (§ 42) we have:

$$\frac{a}{c}+\frac{b}{c}=\frac{a+b}{c};$$

$$\frac{a}{c}-\frac{b}{c}=\frac{a-b}{c};$$

$$\frac{2}{x}-\frac{1}{x}=\frac{1}{x};$$

$$\frac{a}{a-z}-\frac{z}{a-z}=\frac{a-z}{a-z}=1;$$

$$\frac{a+b}{2} + \frac{a-b}{2} = \frac{a+b+a-b}{2} = a \; ;$$

$$\frac{a+b}{2} - \frac{a-b}{2} = \frac{a+b-(a-b)}{2} = b \; ;$$

$$\frac{ac-by}{ac} + \frac{cx}{ac} + \frac{by}{ac} = \frac{ac+cx}{ac} = \frac{c(a+x)}{ac} = \frac{a+x}{a} \; ;$$

$$\frac{a}{aa-yy} + \frac{y}{aa-yy} = \frac{a+y}{(a+y)(a-y)} = \frac{1}{a-y} . \quad \text{(see §§ 91, 92).}$$

If the fractions have not a common denominator, they may be reduced to a common denominator, but the advisibility of doing this depends upon circumstances, as the expression is not always simplified by such reduction.

For example, to reduce the fractions $\frac{a}{b}$ and $\frac{c}{d}$ to a common denominator we multiply the numerator and denominator of the first by d and the second by b. The result is :

$$\frac{a}{b} = \frac{ad}{bd} \quad \text{and} \quad \frac{c}{d} = \frac{bc}{bd}$$

EXAMPLES :

$$\frac{a}{b} + \frac{c}{d} = \frac{ad}{bd} + \frac{bc}{bd} = \frac{ab+bc}{bd} \; ;$$

$$\frac{a}{b} + c = \frac{a}{b} + \frac{bc}{b} = \frac{a+bc}{b} \; ; \quad \text{(see § 34)}$$

$$\frac{12-a+b}{2} + a = \frac{12-a+b+2a}{2} = \frac{12+a+b}{2} \; ;$$

$$\frac{8a+6b}{4} - \frac{6a-2b}{3} = \frac{3(8a+6b)-4(6a-2b)}{12} = \frac{13b}{6} \; ;$$

$$\frac{a}{b} + \frac{a+b}{cd} = \frac{acd+b(a+b)}{bcd} \; ;$$

$$\frac{a}{b} + \frac{a+b}{cd} + \frac{aa-bb-ab}{bcd} = \frac{acd+ab+bb+aa-bb-ab}{bcd} = \frac{a(cd+a)}{bcd} \; ;$$

$$a - \frac{az}{z+1} = \frac{a(z+1)}{z+1} - \frac{az}{z+1} = \frac{a}{z+1} \; ;$$

$$a - \frac{am}{m+n} = \frac{am+an-am}{m+n} = \frac{an}{m+n} \; ;$$

Multiplication and Division of Fractions.

94.

(2) In multiplying two fractions together, we proceed as with numbers, multiplying numerator by numerator and denominator by denominator. If one fraction is to be divided by another, we invert the divisor and proceed as in multiplication. (See §§ 44, 47.)

EXAMPLES :

$$\frac{a}{b} \cdot \frac{c}{d} = \frac{ac}{bd} ; \qquad \frac{a}{b} \cdot c = \frac{ac}{b} ; \qquad \frac{a}{b} : \frac{c}{d} = \frac{ad}{bc} ;$$

$$c : \frac{a}{b} = \frac{bc}{a} ; \qquad 8mn \cdot \frac{3xy}{4my} = 6nx ; \qquad \frac{1}{x} \cdot \frac{6x}{z} \cdot \frac{z}{3x} = \frac{2}{x} ;$$

$$\frac{5am}{bc} : \frac{15mxx}{6ac} = \frac{5am}{bc} \cdot \frac{6ac}{15mxx} = \frac{2aa}{bxx} ;$$

$$\frac{2ax}{3} : \frac{2ax}{3bc} = \frac{2ax}{3} \cdot \frac{3bc}{2ax} = bc ;$$

$$\frac{n}{m} : \frac{ma}{m+n+p} = \frac{na}{m+n+p} ;$$

$$\frac{a+x}{a-x} : \frac{ax+xx}{ax-xx} = \frac{a+x}{a-x} \cdot \frac{x(a-x)}{x(a+x)} = 1.$$

95.

The following expressions in letter symbols are to be reduced by the student to their simpler forms. Every operation should be carefully and conscientiously performed, as the practice thus gained will be found most important in connection with the subjects which follow.

EXAMPLES :

(1) $a+2b-5+a$;   (2) $a-(+b)$;

(3) $a-(-b)$;   (4) $-a \cdot -b$;

(5) $-a \cdot -1$;   (6) $a \cdot \dfrac{b}{2} \cdot 4$;

(7) $3a \cdot 4b \cdot \tfrac{2}{3}c$;   (8) $\tfrac{3}{4}ax \cdot \tfrac{2}{3}by$;

(9) $\tfrac{1}{4}tz \cdot 3x \cdot \tfrac{2}{3}$;   (10) $\dfrac{a}{b}$

(11) $\dfrac{-a}{b}$;

(12) $\dfrac{a}{-b}$;

(13) $\dfrac{-a}{-b}$;

(14) $\dfrac{x}{x}$;

(15) $\dfrac{-x}{x}$;

(16) $\dfrac{2a}{2b}$;

(17) $\dfrac{2x}{x}$;

(18) $\dfrac{x}{2x}$;

(19) $\dfrac{3axyy}{9axy}$;

(20) $\dfrac{6abb}{4bc}$;

(21) $\dfrac{a+y}{a+y}$;

(22) $\dfrac{a+y}{a-y}$;

(23) $\dfrac{\frac{2}{3}abb}{\frac{3}{4}aab}$;

(24) $\dfrac{a}{x}+\dfrac{b}{x}$;

(25) $ax+6ax-8ax+5bx$;

(26) $\frac{1}{3}ay-\frac{2}{3}ay+ay$;

(27) $2ax-5by-16-(ax-16+5by)$;

(28) $(a+b)(a+b)$;

(29) $\frac{2}{3}alx-\frac{2}{3}bzz-(-3alx+\frac{2}{3}bzz)$;

(30) $(ax+by)(ax-by)$;

(31) $(a+b+c)(a+b-c)$;

(32) $\left(\dfrac{x}{2}+\dfrac{y}{3}\right)\left(\dfrac{x}{2}-\dfrac{y}{3}\right)$;

(33) $6aab-12abc+3ab$;

(34) $x+mx+nx+px$;

(35) $\dfrac{m}{2nx}\cdot\dfrac{n}{3m}$;

(36) $\dfrac{a+x}{a-x}\cdot\dfrac{a-x}{a+x}$;

(37) $\dfrac{aa-bb}{a-b}$;

(38) $\dfrac{aa-bb}{a+b}$;

(39) $\dfrac{ay+xy}{3abx+3bxx}$;

(40) $\dfrac{2ay-5by}{4axy-10bxy}$;

(41) $\dfrac{a+x}{3}+\dfrac{a+x}{3}$;

(42) $\dfrac{a}{b}:\dfrac{1}{c}$;

(43) $\dfrac{a}{x}:\dfrac{2x}{y}$;

(44) $7ax:\dfrac{14ax}{5by}$;

(45) $\dfrac{a-x}{a+x}:\dfrac{aa-xx}{ax+xx}$;

(46) $\dfrac{a(b-c)}{2}-\dfrac{b(a-c)}{2}$;

(47) $\dfrac{100a+ap}{100}:\left(1+\dfrac{p}{100}\right)$;

(48) $\dfrac{az+bc}{az}-\dfrac{bc+xz}{az}$;

(49) $\dfrac{16a-5b}{4b} - \dfrac{12a-2b}{3b}$; (50) $\dfrac{2x}{2y-c}\cdot\left(\dfrac{y+c}{3}-\dfrac{c}{2}\right)$;

(51) $\dfrac{x}{a-\dfrac{ac}{x+c}}$; (52) $\dfrac{1-\dfrac{aa}{xx}}{\dfrac{1}{x}-\dfrac{a}{xx}}$;

(53) $\dfrac{a+\dfrac{x-a}{1+ax}}{1-\dfrac{a(x-a)}{1+ax}}$;

(54) $\left(\left[a-\dfrac{m(bn-a)}{n-m}\right]\cdot\dfrac{n-m}{n}+mb\right):\dfrac{a}{x}$;

(55) $\left(\left[\dfrac{1+\dfrac{aa-xx}{aa+xx}}{1-\dfrac{aa-xx}{aa+xx}}+1\right]\cdot\dfrac{1}{1+\dfrac{aa}{xx}}+\dfrac{aa-xx}{a-x}\right)\cdot\dfrac{a}{1+a+x}$;

(56) $\left(\dfrac{\left[\dfrac{(a-1)(ax-bx)}{x-ax}+c\right](a-x)}{a(b+c-a)}-\dfrac{a-x}{a}\right)\cdot\dfrac{aa+xx}{2(aa-xx)}$.

Answers :

(1) $2(a+b)-5$; (2) $a-b$;

(3) $a+b$; (4) ab ;

(5) a ; (6) $2ab$;

(7) $8abc$; (8) $\dfrac{abxy}{2}$;

(9) $2txz$; (10) $\dfrac{a}{b}$;

(11) $-\dfrac{a}{b}$; (12) $-\dfrac{a}{b}$;

(13) $\dfrac{a}{b}$; (14) 1 ;

(15) -1 ; (16) $\dfrac{a}{b}$;

(17) 2 ; (18) $\frac{1}{2}$;

(19) $\dfrac{y}{3}$;

(20) $\dfrac{3ab}{2c}$;

(21) 1 ;

(22) $\dfrac{a+y}{a-y}$;

(23) $\dfrac{8b}{9a}$;

(24) $\dfrac{a+b}{x}$;

(25) $5bx-ax=(5b-a)\,x$;

(26) $\dfrac{2}{3}ay$;

(27) $ax-10by$;

(28) $aa+2ab+bb$;

(29) $\dfrac{11alx}{3}-\dfrac{19bzz}{15}$;

(30) $aaxx-bbyy$;

(31) $aa+2ab+bb-cc$;

(32) $\dfrac{xx}{4}-\dfrac{yy}{9}$;

(33) $3ab\,(2a-4c+1)$;

(34) $(1+m+n+p)x$;

(35) $\dfrac{1}{6x}$;

(36) 1 ;

(37) $a+b$;

(38) $a-b$;

(39) $\dfrac{y}{3bx}$;

(40) $\dfrac{1}{2x}$;

(41) $\dfrac{2}{3}(a+x)$;

(42) $\dfrac{ac}{b}$;

(43) $\dfrac{ay}{2xx}$;

(44) $\dfrac{5by}{2}$;

(45) $\dfrac{x}{a+x}$;

(46) $\dfrac{c\,(b-a)}{2}$;

(47) a ;

(48) $\dfrac{a-x}{a}$;

(49) $-\dfrac{7}{12}$;

(50) $\dfrac{x}{3}$;

(51) $\dfrac{x+c}{a}$;

(51) $a+x$

(53) x ;

(54) x ;

(55) a ;

(56) 1.

BOOK XI.

Algebra.

Simple Equations with One Unknown Quantity.

96.

Nearly all mathematical investigations lead into the investigation of problems which require for their final solution the help of arithmetic, and involve not only a thorough knowledge of arithmetic but also a certain degree of mechanical dexterity. In order to carry the study of arithmetic beyond the limits of the first part of this work, it is necessary not only to develop the theory, but also to be able to grasp the practical questions, without which a further development of the subject is impracticable. The first portion of this work dealt only with the theory of number systems, with the first four rules, both for whole and fractional numbers, and with the rules of proportion ; together with a few of the simplest applications of these theories.

Now as a matter of fact all calculations, no matter how complicated they may appear, may be carried back to the first four rules of arithmetic. But before we can bring the many conditions of a problem into their final form, there must often be many statements and deductions made, and the four simple rules must be turned and twisted, combined and separated in many ways. For these reasons it is necessary, in order to keep the premises and conclusions clearly in mind, to use all the methods of abridgment and simplifying symbols possible, and thus to avoid confusion wherever practicable. Especially is it important to study first the so-called "simple" equations, and learn to apply them to the solution of practical problems, remembering that after all this is nothing but a symbolical application of the rule of three.

The student cannot be too thorough in studying this subject, upon which so many others depend, and should make sure that he understands the following simple, but very important matters.

97.

Two expressions of the same value can be placed *equal* to each other. Thus 18—4 is just as much as 6+3+5, and so we can write :

$$18-4=6+3+5$$

and such a statement of equality, as already explained in §16, is called an *equation*. All that stands on the right of the sign of equality ($=$) is called the right hand side of the equation, all on the left, the left hand side; and the various quantities (letters, figures, &c.), which make up the two sides, are called the *members* of these sides. For example, in the above equation, $+5$ is the third member of the right side, and -4, is the second member of the left side.

98.

If we transpose (or transfer) any member from one side of an equation to the other side, at the same time changing its *sign*, the equation will still be true. Thus, for example, in the equation :

$$18-4=6+3+5 \qquad (1)$$

if we take the second member of the left side and, *changing its sign,* place it over on the right side, we have the following new equation :

$$18=6+3+5+4 \qquad (2)$$

The truth of this simple, but very important, fact, follows from the general truth, "Equals added to, or subtracted from, equals, give equals." (See § 18). The value of each of the two sides of equation (1) is equal to 14. If we add $+4$ to *both sides*, they must still remain equal *to each other*. The left side will now be 18—4+4, or 18, since the —4 and +4 cancel each other. Briefly, then, the transposition of any member of an equation from one side to the other, changing its sign at the same time, is equivalent to adding or subtracting the same member from both sides of the equation.

Suppose we take equation (2) and transpose the first two members of the right side to the left side, also changing their signs, we have the true equation,

$$18—6—3=5+4 \qquad (3)$$

this being just the same as if we had subtracted—6 and 3 from both sides. The value of each side is now 9.

If we transpose the first member of the left side to the right, we have :

$$—6—3=5+4—18 \qquad (4)$$

and the algebraic sum of both sides= —9.

From (4), by transposing the two members from the left side to the right, we have :

$$0=5+4—18+6+3 \qquad (5)$$

If we change *all* the signs on *both* sides the equation will still remain true. Thus, for instance, from :

$$18—4=6+3+5 \qquad (6)$$

by changing all the + to —, and all the — to +, we get :

$$—18+4= —6—3—5 \qquad (7)$$

which is evidently true, since the values of the two sides are not changed but only reversed in sign. The value of each side of (6) is 14, of each side of (7) is —14. (See § 76).

99.

We have seen that by transposing members from either side of an equation the truth of the equation is not affected, and hence if we reduce one side of an equation to a single member by transposing all the other members to the other side, the value of all the members on this last side will be equal to that of the one member on the first side. Thus if we take the equation :

$$6+5—3=10—2 \qquad (1)$$

and reduce the left side to +5 by transposing the other members +6 and —3 to the right hand side, we have :

$$5=10—2—6+3 \qquad (2)$$

and the value of the right hand side must be equal to 5. (§ 98).

Let us now suppose that in equation (1) there was one member whose value was *unknown*, and in its place any symbol, such as x, be placed, and let it be required to find the value of x.

Suppose the equation to be :

$$6+x-3=10-2 \qquad (1)$$

To *solve* the equation, means to find a number, which when put in place of x, will still keep the values of both sides equal to each other. The equation is then said to be *satisfied*.

This we may do by transposing all the members but x to the right hand side, when we have :

$$x=10-2+3-6$$

and collecting the right hand members (according to § 76) we have :

$$x=5$$

100.

We see that to *solve* an equation is to find a value for the unknown quantity x, which shall make the left side equal to the right. The study of the solution of equations is called *Algebra*. The equation $4+x=13$, for example, is *solved* when it is reduced to the form $x=9$, since $4+9=13$. The value for the unknown quantity (9 in the foregoing example) is called the *root* of the equation. Since the unknown quantity may be combined with the known or given quantities in a variety of ways, we must first classify the various forms, and explain simple examples under each form before we can proceed to the application of equations in solving algebraic problems.

101.

First Case. When the unknown quantity appears in only one member of the equation, with a *factor*, or *coefficient*.

Reduce one side of the equation to a single member, containing the unknown quantity, then collect the members of the other side into one by taking their algebraic sum, and divide this sum by the coefficient of the unknown quantity.

Suppose in the equation $4+7\cdot8=60$, the quantity 8 was unknown, and instead the symbol x was written, thus :

$$4+7x=60$$

We first transpose the $+4$,

$$7x=60-4$$

Collecting, $\qquad 7x=56$

Dividing by 7, $\qquad x=8$

It is quite clear that if $7x=56$, x must equal $\frac{1}{7}$ as much, i. e. $\frac{56}{7}=8$.

EXAMPLE : If we subtract 5 from six times a certain number the remainder will be 14. What is the number?

Solution : If we call the unknown number x, then six times the number will be $6x$, and we have from the question the following equation :

$$6x-5=14$$
Transposing, $6x=14+5$
Collecting, $\quad 6x=19$

Dividing by the coefficient of x : $\quad x=\dfrac{19}{6}=3\dfrac{1}{6}$.

102.

Second Case. When the unknown quantity appears in only one member of the equation, with a *divisor*.

In this case reduce one side to a single member containing the unknown quantity, and then multiply both sides by the divisor.

Suppose in the equation $\frac{18}{6}+4=7$, the quantity 18 was unknown, and instead the symbol x, was written, we should have :

$$\frac{x}{6}+4=7$$
Transposing, $\frac{x}{6}=7-4$
Collecting, $\frac{x}{6}=3$
Multiplying by 6, $x=18$

Since the sixth part of $x=3$, x itself must be 6 times as great.

EXAMPLE. If we divide a certain number by 5 and subtract $\frac{2}{3}$ from the quotient, the result will be $\frac{3}{7}$, what is the number?

Solution. Placing the unknown $=x$, and stating the question as an equation, we have :

$$\frac{x}{5}-\frac{2}{3}=\frac{3}{7}$$
Transposing, $\frac{x}{5}=\frac{3}{7}+\frac{2}{3}$
Collecting, $\frac{x}{5}=\frac{23}{21}$
Multiplying by 5, $x=\frac{115}{21}$

If the unknown on the left side should be negative, we multiply both sides by —1. Thus, for example:

$$3x+5=4x$$
$$3x-4x=5$$
$$-x=-5$$

Multiplying by —1, $x=5$

103.

Third Case. When the unknown appears in only one member of the equation, but with both a factor and a divisor.

Reduce the equation to the unknown member, as before, aud then divide the value on the other side by the fractional coefficient; i. e., multiply by the divisor and divide by the factor. (See § 47.)

Suppose, in the equation $6+\frac{4\cdot10}{5}=14$, we had x instead of 10, the equation would become:

$$\frac{4x}{5}=8$$
whence $x=8\cdot\frac{5}{4}$
$$x=10$$

for if $\frac{4}{5}$ of a quantity is equal to 8, it is clear that the quantity itself is equal to $\frac{5}{4}\times8=10$.

Or we may put it, if the 5th part of $4x$ is equal to 8, clearly $4x=8\cdot5$, and $x=\frac{8\cdot5}{4}=10$.

EXAMPLE. There is a certain number which if multiplied by 7, and the product divided by 9, and then $\frac{2}{5}$ added to the quotient, the result will be $\frac{6}{11}$. What is the number?

Solution. Stating the question in the form of an equation, we have:

$$\frac{7x}{9}+\frac{2}{5}=\frac{6}{11}$$
whence $\frac{7x}{9}=\frac{6}{11}-\frac{2}{5}$
or $\frac{7x}{9}=\frac{8}{55}$
$$x=\frac{8}{55}\cdot\frac{9}{7}$$
$$x=\frac{72}{385}$$

104.

Fourth Case. When the unknown quantity appears in several members of an equation. In this case separate all the known from the unknown members, by transposing all members containing the unknown quantity to the left side of the equation, and all containing known quantities to the right hand side.

Then collect each side into a single member, when the solution will be the same as in the third case.

If in the equation $4+3.5=49-4.5-2.5$, the number 5 is unknown, and the symbol x written for it, we have:

$$4+3x=49-4x-2x$$

Transposing all members containing x to the left side, and all known members to the right side, we have:

$$3x+4x+2x=49-4$$
$$\text{Collecting,} \quad 9x=45$$
$$\text{Dividing by 9,} \quad x=5$$

It is clear that a quantity multiplied by 3, by 4, and by 2 is the same as if it were at once multiplied by 9.

EXAMPLE. What number is it which when multiplied by 10 gives the same result as when 10 is added to it?

Solution. Let x be the number, then we have:

$$10x=x+10$$
$$\text{Transposing,} \quad 10x-x=10$$
$$\text{Collecting,} \quad 9x=10$$
$$\text{Dividing by 9.} \quad x=1\tfrac{1}{9}$$

105.

Fifth Case. When an unknown quantity stands in a parenthesis with known quantities. In this case we first proceed to remove the parenthesis by multiplying the quantities within by the factor which stands without the parenthesis. (See §19.)

Suppose, in the equation $6\cdot4=9+3\cdot5$, we have instead of the single factors 4 and 5, the binomial factors $7-3$ and $12-7$, which (according to §19) must be enclosed in parentheses. The equation would then read:

$$6(7-3)=9+3(12-7)$$

Now suppose the number 7 to be the unknown quantity, indicated by x, the equation is:

$$6(x-3)=9+3(12-x) \qquad (1)$$

from which we must determine x.

In order to get the unknown quantity out of the parentheses we must (according to §19) multiply every quantity within the parenthesis by the factor without, which gives :

$$6x-18=9+36-3x \qquad (2)$$

whence, $\quad 6x+3x=9+36+18$

and, $\qquad 9x=63$

hence, $\qquad x=7$

106.

In problems involving the use of parentheses we must be careful to consider those which are preceded by a minus sign.

Take, for example, the following equation :

$$6(x-3)-3(12-x)=9 \qquad (3)$$

and we have according to § 89, Case 2,

$$6x-18-36+3x=9 \qquad (4)$$
$$6x+3x=9+18+36$$
$$9x=63$$
$$x=7$$

EXAMPLE. What number is it which, when we subtract 6 from it, multiply the remainder by 5 and subtract this product from 10, will equal zero ?

Solution. Let the required number be x. We then have for the first remainder, $x-6$, and this multiplied by 5, gives $5(x-6)$. Substracting this product from 10 gives $10-5(x-6)$ and since this is equal to zero, we have :

$$10-5(x-6)=0 \qquad (1)$$

whence $\quad 10-5x+30=0 \qquad (2)$

$$-5x=-30-10 \qquad (3)$$
$$-5x=-40 \qquad (4)$$

changing signs on both sides, according to § 98, we have :

$$5x=40$$
$$x=8$$

Instead of changing the signs in (4), we might have transposed the unknown to the other side, or, we should have obtained the result more simply by dividing both sides of (4) at once by -5.

107.

Finally we have the case in which the unknown exists in several members with a divisor. Suppose in the equation

$2 \cdot 24 + \frac{7 \cdot 24}{6} - 22 = 40 - \frac{2 \cdot 24}{3} + \frac{5 \cdot 24}{4}$ the factor 24 is the unknown quantity.

Replacing it by x we have:

$$2x + \tfrac{7x}{6} - 22 = 40 - \tfrac{2x}{3} + \tfrac{5x}{4} \qquad (1)$$

transposing all members containing x to the left side:

$$2x + \tfrac{7x}{6} + \tfrac{2x}{3} - \tfrac{5x}{4} = 40 + 22 \qquad (2)$$

Now before we can collect the members containing x, we must reduce the fractional coefficients to a common denominator, and thus we get:

$$2 + \tfrac{7}{6} + \tfrac{2}{3} - \tfrac{5}{4} = 2 + \tfrac{14}{12} + \tfrac{8}{12} - \tfrac{15}{12} = 2\tfrac{7}{12}$$

and therefore

$$2\tfrac{7}{12}x = 62$$
$$\tfrac{31x}{12} = 62$$
$$x = 24$$

Instead of working the equation (2) in the above manner it is, however, generally found simpler, first to collect the quantities on the right side, and then multiply both sides by the least common denominator, determining this by the method already given in §23. Thus we have from equation (2):

$$2x + \tfrac{7x}{6} + \tfrac{2x}{3} - \tfrac{5x}{4} = 62 \qquad (3)$$

then multiplying both sides by 12 all the denominators will disappear, since they will all divide 12 without remainder, we get:

$$12 \cdot 2x + 12 \cdot \tfrac{7x}{6} + 12 \cdot \tfrac{2x}{3} - 12 \cdot \tfrac{5x}{4} = 62 \cdot 12$$

or more briefly (see §23):

$$24x + 14x + 8x - 15x = 62 \cdot 12$$
$$31x = 62 \cdot 12$$
$$x = \tfrac{62 \cdot 12}{31}$$
$$x = 24$$

EXAMPLE. Required a number such that when $\frac{2}{3}$ of itself be subtracted from $\frac{3}{4}$ of itself the remainder will equal 14.

Solution. Let x be the number; then from the above conditions we have:

$$\tfrac{3}{4}x - \tfrac{2}{3}x = 14$$

then multiplying both sides by $4 \cdot 5$ we have:

$$15x - 8x = 280$$
$$7x = 280$$
$$x = 40$$

Example. Required a number such that when we subtract 15 from 6 times the number, multiply the remainder by $\frac{4}{5}$ and then subtract the product from 8 times the number, the result will equal 12.

Solution. Calling the number x, we have :

$$8x - \tfrac{4}{5}(6x - 15) = 12$$

multiplying both sides by 5, we have :

$$40x - 4(6x - 15) = 60$$
$$40x - 24x + 60 = 60$$
$$16x = 0$$
$$x = \tfrac{0}{16} = 0$$

BOOK XII.

Applications of Algebra.

Solution of Algebraic Problems with One Unknown Quantity.

108.

Algebraic problems are of such an infinite variety that, unlike ordinary arithmetical examples, they do not admit of a close classification. Many varied operations must be performed not only upon unknown quantities, but also upon symbols standing for unknown quantities (usually the last letters of the alphabet, x, y, z, t, etc.,) and while these quantities remain as yet unknown, the operations can only be indicated.

General rules for the solution of such problems cannot be given, and it is only by careful and intelligent consideration and thought that each special problem can be framed in words and stated in algebraic symbols. The known and unknown quantities must be so grouped and framed together that we can construct two expressions which are equal to each other, and then placing these on opposite sides of the sign of equality ($=$) we have framed our *equation*.

It is in this matter of *stating* the problem that the real ability of the student must show itself. In the text books there are many examples given, which by careful study he may learn how to solve. In practical life, however, the problems which the engineer, the physicist, the electrician, and the business man, are all called to solve, are not mere examples stated in the convenient terms of the text book, but are mingled with a number of distracting conditions and surroundings which often require keen observation and clear judgment to disentangle them before a problem can be placed in mathematical form at all.

When the equation expressing the relations between the known and unknown quantities is once stated the rules of the preceding chapter may readily be applied.

The matter of the statement of the equation, however, is a question of judgment and acute reasoning. These faculties cannot be created by study, but can be greatly developed and cultivated by exercise and use.

Here, as in all actual applications of mathematics, the mathematician must depend upon himself. He himself must be the judge of the validity of his conclusions, and separate the true from the false.

Here it is that the beginner, who has not yet acquired the use of his own powers, nor learned to think and act for himself, must expect to find his greatest difficulties, but he must not be deterred or frightened off by any fears or lack of self-confidence. Practice and industry will soon develop a skill and facility of judgment which will render the work less and less difficult, and give constantly increasing confidence to the student in his work. Above all he must remember that success does not depend upon the *number* of examples which he may work out. A single problem carefully thought out and brought to a successful conclusion by one's own unaided efforts, is worth a thousand hurried through in the same time with assistance from others. The examples here given are intended to show the way and point out the methods. The student will do well, however, after having carefully read them through, to work them out independently a second time without reference to the book at all.

All mathematics consists of problems. The actual algebraic work, however, is by far the easier portion, since it requires little or no judgment, but only the exercise of care and accuracy, as is shown by the fact that many of the more complicated mathematical operations can readily be performed by machinery. No machine, however, can ever be made to take the place of human judgment, and the training which is given to the mental faculties in the study of mathematics, will be found to strengthen the mind most effectively for all purposes and in all walks of life.

The problems which are merely stated, and involve the application of the rules of algebra, are only intended for beginners, but it is necessary to give some time at first to them, since he who cannot perform the simple operations involved in their solution, can hardly expect to make rapid progress in mathematics.

109.

EXAMPLE 1. There is a certain number, which when multiplied by 2, and added to itself multiplied by 3, is equal to itself multiplied by 7 subtracted from 36. What number is it?

Solution. Call the unknown number x, then $2x$ is twice the number, $3x$ three times the number, and $7x$ is seven times the number. According to the question, we have $2x+3x$ is equal to $7x$ subtracted from 36, so we may state the equation :

$$2x+3x=36-7x$$

Transposing $7x$ to the left side,

$$2x+3x+7x=36$$
$$\text{Collecting,} \quad 12x=36$$
$$\text{Whence,} \quad x=3$$

In order to prove the correctness of the work we may substitute 3 for x in the original equation, to see if the two sides *are* then equal to each other.

110.

EXAMPLE 2. What number is it, which when multiplied by 10 is equal to itself added to 3?

Solution. Let the number be represented by x, and from the conditions of the question we have :

$$10x=x+3$$
$$10x-x=3$$
$$9x=3$$
$$x=\tfrac{3}{9}=\tfrac{1}{3}$$

111.

EXAMPLE 3. Three persons, A, B. and C, have \$36 to divide amongst them in such a proportion that B shall have twice as much as A, and C three times as much as B. How much should each one receive?

Solution. Let x represent the amount which A receives. Then B's share will be $2x$, and C's share $6x$, and since all three shares must equal \$36, we have the equation:

$$x+2x+6x=36$$
$$\text{or,} \quad 9x=36$$
$$x=4$$

hence A's share$= x=4$

B's share$=2x=8$

C's share$=6x=24$

$$\text{Total,} \quad \overline{36}$$

112.

EXAMPLE 4. \$100 is to be divided between four persons, A, B, C, and D, so that their portions shall be in the proportion of 3, 5, 8, and 4. What is the amount of each one's share?

Solution. Let A's share $=x$. The proportions 3, 5, 8 and 4 are the same as 1, $\frac{5}{3}$, $\frac{8}{3}$ and $\frac{4}{3}$ (§65), and so we have, B's share $=\frac{5}{3}x$, C's share $=\frac{8}{3}x$, and D's share $=\frac{4}{3}x$, and hence:

$$x+\tfrac{5x}{3}+\tfrac{8x}{3}+\tfrac{4x}{3}=100$$
$$\text{or,} \quad \tfrac{20x}{3}=100$$
$$\text{and,} \quad 20x=300$$
$$x=15, \text{ A's share.}$$
$$\tfrac{5}{3}x=25, \text{ B's share.}$$
$$\tfrac{8}{3}x=40, \text{ C's share.}$$
$$\tfrac{4}{3}x=20, \text{ D's share.}$$

113.

EXAMPLE 5. \$6,000 is to be divided among three persons, A, B, and C, in such a manner that B is to receive three times as much as A, less \$200; C, four times as much as B, but plus \$200. What is the share of each?

Solution. Let A's share be$=x$, then B's will be $3x-200$, and C's share will be $4(3x-200)+200$. We then have the equation:

$$x+3x-200+4(3x-200)+200=6000$$
$$x+3x+12x-800=6000$$
$$16x=6800$$
$$x=425$$

Hence the shares will be: A, \$425; B, \$1075; and C, \$4,500.

114.

EXAMPLE 6. A Greek went into the temple of Jupiter and prayed to the god that he would double his money for him. Jupiter did so, and in gratitude the man laid 2 oboli upon the altar as a gift. He then went with the remainder to the temple of Apollo, and prayed to him to double this remainder. Here also, his prayer was answered, and he left a gift of four oboli. When the Greek went to count his money he found that notwithstanding the repeated doubling, it was all gone ; how much did he have at first?

Solution.   Let $x=$ the original number of oboli.   Then $2x$ would be the amount after the first doubling, and since he gave 2 oboli, the remainder would be $2x-2$. This remainder when doubled by Apollo, would be $2(2x-2)$, and as he gave up 4 oboli, it made $2(2x-2)-4$. As he then had nothing left, the equation will be :

$$2(2x-2)-4=0$$
$$\text{whence} \quad 4x-4-4=0$$
$$4x=8$$
$$x=2$$

115.

EXAMPLE 7.   Divide the number 100 into two parts, so that the larger part divided by 6, and the smaller part divided by 4 shall give equal quotients.

Solution.   At first it looks as if we had here *two* unknown quantities.   If we knew one of them, however, we should know the other, since the second would be equal to the first, subtracted from 100.   Suppose then, that we call the greater of the two numbers $=x$, we have for the other $100-x$, and thus can express them both without using a second unknown symbol.   We then have from the terms of the question :

$$\frac{x}{6}=\frac{100-x}{4}$$

Multiplying both sides by 12 we get :

$$2x=3(100-x)$$
$$2x=300-3x$$
$$5x=300$$
$$x=60$$

Since $x=60$, the second number is $100-60=40$.

116.

EXAMPLE 8. Required to divide the number 100 into two parts, such that when one is divided by 5, and the other by 3, the sum of the quotients shall be equal to 24.

Solution. Let x equal the first number, and $100-x=$ the second number. ʹThen from the question we have :

$$\frac{x}{5}+\frac{100-x}{3}=24$$

Multiplying by 3·5, we get :

$$3x+500-5x=360$$
$$-2x=-140$$
$$x=70 \quad \text{(See § § 98 and 106)}.$$

hence 70 is one number, and $100-70=30=$ the other.

117.

EXAMPLE 9. A man was asked how old he was, and replied : "When I am as many years over 100 as I now am under 100, I will be just twice as old as I am now." What was his age ?

Solution. Let x be the required age. Then $100-x$ will be the number of years he lacks of being 100 years of age. If he were that many years over 100, he would $100+(100-x)$. But this, according his statement, is just twice his present age, so it must be equal to $2x$, hence the equation is :

$$2x=100+100-x$$
$$3x=200$$
$$x=66\tfrac{2}{3}$$

118.

EXAMPLE 10. Some one asked Pythagoras how many pupils he had. He answered in the following enigmatical manner : Half of them study philosophy ; one-third study mathematics. If to the remainder, who keep silence, you add the three pupils I have with me, and which are not included in the foregoing, you will have just one-fourth of those who study mathematics and philosophy." How many pupils had he ?

Solution. Let x be the number of pupils, then $\frac{x}{2}$ study philosophy, and $\frac{x}{3}$ study mathematics. These together make $\frac{1}{2}x+\frac{1}{3}x=\frac{5}{6}x$, so that the remainder must equal $\frac{1}{6}x$. But if 3 be added to this remainder, it will equal $\frac{1}{4}$ of the number who study philosophy and mathematics, i. e. $\frac{1}{4}$ of $\frac{5x}{6}$, so we have for the equation :

$$\frac{x}{6}+3=\frac{1}{4}\cdot\frac{5}{6}x$$

$$\frac{x}{6}-\frac{5x}{24}=-3$$

Multiplying by 24 gives

$$4x-5x=-72$$

$$-x=-72$$

or, multiplying both sides by —1,

$$x=72$$

119.

EXAMPLE 11. A builder can build a certain wall in 6 days, and another can build it in 3 days. What time will it take them to build it if both work together?

Solution. The man who can do the work in 6 days must do $\frac{1}{6}$ of it in 1 day, while the second man can do $\frac{1}{3}$ of it in 1 day. Both working together will then do $\frac{1}{6}+\frac{1}{3}=\frac{9}{18}$ of the work in 1 day. If then we make $x =$ the number of days required to do the whole work ($=1$), we have:

$$\frac{9}{18}\cdot x=1$$

$$x=2$$

120.

EXAMPLE 12. A certain pool is to be pumped dry. One pumping engine is capable of removing the water in 30 days, another one can do it in 40 days, and a third can pump it out in 20 days. How many days will be required if all three pumps work together?

Solution. Let us call the quantity of water $=1$. The first pump can remove this water in 30 days, hence $\frac{1}{30}$th of it in 1 day. In like manner the daily portion of the second is $\frac{1}{40}$th and the third pump $\frac{1}{20}$th. All three pumps would then remove $\frac{1}{30}+\frac{1}{40}+\frac{1}{20}=\frac{13}{120}$ of the amount in one day. In x days the amount pumped would be $\frac{13}{120}x$, whence:

$$\frac{13x}{120}=1$$

$$x=\frac{120}{13}=9\frac{3}{13}$$

121.

EXAMPLE 13. A number of apples were to be divided among some children. In order that each child should have five apples, two more apples would have been required. When each child received four apples, there were three apples left over. How many apples, and how many children were there?

Solution. This question appears to have two unknown quantities, but one of them can be obtained from the other. Suppose we call the number of children$=x$. If each were to have 5 apples, there would be two apples short; whence $5x-2=$ the number of apples. If each child had 4, there would be 3 apples over, hence $4x+3$ also$=$ the number of apples. These two expressions must then be equal to each other, hence:

$$5x-2=4x+3$$
$$5x-4x=5$$
$$x=5$$

There were then 5 children, and $5 \cdot 5-2=4 \cdot 5+3=23$ apples.

122.

EXAMPLE 14. In a certain assemblage, there were 3 times as many men as women. Later, 8 men with their wives left, and there were then in the remaining company, 5 times as many men as women. How many men, and how many women were there present in the first instance?

Solution. If we call the original number of women $=x$; the original number of men will be $3x$; when 8 men and 8 women have gone, there will be remaining $x-8$ women, and $3x-8$ men; and since now the men are 5 times as many as the women, we have:

$$3x-8=5 (x-8)$$
$$\text{whence,} \quad -2x=-32$$
$$x=16$$

16 women and $3 \times 16=48$ men.

123.

EXAMPLE 15. A servant was to receive $135 a year and a suit of clothes. He left the situation after working 7 months, and received as his pay $67 besides the clothes. How much was allowed as the value of the clothes?

Solution. Let the value of the clothes be $=x$. Then the pay for a year would be $135+x$, and for one month$= \frac{135+x}{12}$, and for 7 months $=\frac{7}{12}(135+x)$. But this amount will also be equal to \$67$+x$, whence we have :

$$67+x=\tfrac{7}{12}(135+x)$$

multiplying by 12,

$$804+12x=945+7x$$
$$5x=141$$
$$x=28\tfrac{1}{5}$$

Remember that all the members of an equation must be of one and the same denomination.

124.

EXAMPLE 16. An employer took an apprentice, and they agreed that for every day the apprentice worked he was to receive one dollar, but for every day he was absent he should pay a fine of 60 cents. After 80 days, they reckoned up their accounts and found that neither owed the other anything. How many days had the apprentice been at work?

Solution. Let x be the required number of days. Then 80—x will be the number of days the apprentice was absent. His fine during that time would amount to $\frac{60}{100}(80-x)$, and his earnings during x days would be x dollars, hence :

$$x-\frac{60\,(80-x)}{100}=0$$
$$10x-480+6x=0$$
$$16x=480$$
$$x=30$$

125.

EXAMPLE 17. A merchant bought 120 yards of cloth for \$300. The seller, however, allowed him a discount from this \$300, equal to the net cost of 20 yards to him. What was the amount of the discount in money?

Solution. Let the amount of the discount be $=x$. Then the 120 yards cost 300—x, and 20 yards cost $20\frac{300-x}{120}=x$, and since this is equal to x also, we have :

$$20\tfrac{300-x}{120}=x$$

or

$$x=\tfrac{300-x}{6}$$
$$6x=300-x$$
$$7x=300$$
$$x=42\tfrac{6}{7}$$

126.

EXAMPLE 18. What number is it which has the following properties? When it is multiplied by 3, and 11 subtracted from the product, the remainder then multiplied by 4, and then 6 added to the product, and the sum multiplied by 5, the result will equal 50.

Solution. In order to indicate all these operations by signs, two parentheses will be required. If we call x the number, then we have $3x-11$, and $4(3x-11)+6$ will be the sum. Now to indicate that this is to be multiplied by 5, we enclose it all between parentheses again, and to avoid confusion, these latter are made in the form of brackets, thus: $5[4(3x-11)+6]=50$.

Now to relieve the unknown quantity from the parentheses, we remove first the inner and then the outer parenthesis (or *vice versa*). We will begin with the inner one and obtain first:

$$5[12x-44+6]=50$$
$$\text{or} \quad 5[12x-38]=50$$

Dividing both sides by 5 we have:

$$12x-38=10$$
$$12x=48$$
$$x=4$$

127.

EXAMPLE 19. A market woman brought a basket of apples to market. She sold first one-half of them and half an apple. She next sold one-half of the remainder and half an apple; and then she sold one-half of what was left and half an apple. She then had 24 apples left. How many had she at first?

Solution. Let x be the number of apples. After the first sale she had left $\frac{1}{2}x-\frac{1}{2}$; after the second sale $\frac{1}{2}(\frac{1}{2}x-\frac{1}{2})-\frac{1}{2}$; after the third sale she had left $\frac{1}{2}[\frac{1}{2}(\frac{1}{2}x-\frac{1}{2})-\frac{1}{2}]-\frac{1}{2}$, and this must be equal to the 24 apples which were left, so we have:

$$\frac{1}{2}[\frac{1}{2}(\frac{1}{2}x-\frac{1}{2})-\frac{1}{2}]-\frac{1}{2}=24$$

Removing the inner parenthesis, we have:

$$\frac{1}{2}[\frac{1}{4}x-\frac{1}{4}-\frac{1}{2}]-\frac{1}{2}=24$$
$$\text{or,} \quad \frac{1}{2}[\frac{1}{4}x-\frac{3}{4}]-\frac{1}{2}=24$$

Then removing the outer parenthesis,

$$\frac{1}{8}x-\frac{3}{8}-\frac{1}{2}=24$$
$$\frac{x}{8}=24\frac{7}{8}$$
$$x=199$$

128.

EXAMPLE 20. How great must the principal be, which at 5 per cent. interest, brings in as much in 4 years, as \$635 at 4%, brings in 7 years?

Solution. The money at 4 per cent. brings in \$4 a year for every \$100, hence $\frac{635}{100} \cdot 4$ is the yearly interest and $\frac{635}{100} \cdot 4 \cdot 7$ the total interest of the second investment. Now let $x =$ the principal at 5 per cent., then $\frac{x}{100} \cdot 5 \cdot 4$ must be the total interest of the first investment. But these two are equal to each other, so we have:

$$\frac{x}{100} \cdot 5 \cdot 4 = \frac{635}{100} \cdot 4 \cdot 7$$
$$\text{whence,} \quad x = \$889$$

129.

EXAMPLE 21. What must be the rate of interest for \$225 to bring in as much in 6 years, as \$300 at $3\frac{1}{2}$% will yield in 8 years?

Solution. Let x be the required rate, and we have

$$\frac{225}{100} x \cdot 6 = \frac{300}{100} \cdot \frac{7}{2} \cdot 8$$
$$\text{whence,} \quad 45 x = 10 \cdot 7 \cdot 4$$
$$9 x = 56$$
$$x = 6\frac{2}{9}\%$$

130.

EXALPLE 22. What must be the principal of an investment, which at 4%, for 5 years, makes \$600 total, principal and interest?

Solution. Let x be the required principal. We then have $\frac{x}{100} \cdot 4 \cdot 5$ for the interest at 4% for 5 years, and $x + \frac{x}{100} \cdot 4 \cdot 5$ for the sum of principal and interest;

$$\text{hence,} \quad x + \frac{x}{100} \cdot 4 \cdot 5 = 600$$
$$x + \frac{x}{5} = 600$$
$$5 x + x = 3000$$
$$x = \$500$$

131.

EXAMPLE 23. A desires to send B by express \$5100, but B, wishes the charge of 2% to be prepaid and deducted; how much will B receive?

Solution. Let the required amount be $=x$. The express charges will then be $\frac{x}{100} \cdot 2$, and the total $=x + \frac{x}{100} \cdot 2$. But the total $=\$5100$ hence we have:

$$x + \frac{x}{100} \cdot 2 = 5100$$
$$51 \, x = 5100.50$$
$$x = 5000$$

132.

EXAMPLE 24. A wishes to send B $4900, but cannot prepay the charge of 2% at the office of sending. He desires to make the amount such that B will have $4900 clear after paying the charge of 2%. How much must he send?

Solution. Let x be the required amount, then B must pay on it $\frac{x}{100} \cdot 2$, hence:

$$x - \frac{x}{100} \cdot 2 = 4900.$$
$$49x = 4900 \cdot 50$$
$$x = 5000$$

133.

EXAMPLE 25. A man desires to obtain $30,000 insurance, on which the premium is 20%, but desires to make the amount of the insurance such as to include the amount of the premium also. How much will be the premium?

Let x be the required amount. Then the total amount of the insurance will be $30,000 + x$, and the premium on this will be

$$\frac{30000 + x}{100} \cdot 20$$

But this is equal to x, hence we have:

$$x = \frac{30000 + x}{100} \cdot 20$$
$$4x = 30,000$$
$$x = 7500$$

134.

EXAMPLE 26. A courier who travels 25 miles a day has been dispatched 8 days, when another, traveling 45 miles a day, is sent after him. How many days will it take the second man to overtake the first?

Solution. Let x be the number of days, then the second man will travel a distance of $45x$ miles, while the first, who has had 8 days start, will have traveled $25(x+8)$. At the time the first is overtaken, both have traveled the same distance, hence the two expressions must be equal to each other, and we have :

$$45x=25(x+8)$$
$$45x=25x+200$$
$$20x=200$$
$$x=10$$

135.

EXAMPLE 27. The hour hand of a clock points to VI, and the minute hand to XII. When will the two hands be exactly together?

Solution. Let the time in hours, or fractions of an hour, which will elapse before the hands are together $=x$. The minute hand moves 60 minutes in an hour, and in x hours it will move $60x$ minutes. The hour hand moves over a space equal to 5 minutes in one hour, and in x hours it will move over $5x$ minutes. But it has to move over 30 minutes to reach the place where the minute hand now is. This gives the equation :

$$60x=30+5x$$

for the two expressions must be equal if the two hands are together, whence :

$$x=\tfrac{6}{11} \text{ of an hour}=32\tfrac{8}{11} \text{ minutes}$$

and the hands would be together at $32\tfrac{8}{11}$ minutes after six o'clock.

136.

EXAMPLE 28. Someone being asked what time it was, said the hour and minute hands were exactly together, between X and XI. What time was it?

Solution. The minute hand stood at XII when the hour hand was at X, and was then $10\cdot5=50$ minutes behind the hour hand, hence :

$$60x=50+5x$$
$$x=\tfrac{10}{11}$$

The minute hand was therefore $\tfrac{10}{11}$ minutes before the XII mark, and as the hour hand was at the same place the time was $5\tfrac{10}{11}$ minutes before XI.

137.

EXAMPLE 29. A dog was chasing a rabbit. The rabbit was 100 leaps ahead of the dog and made six leaps while the dog made 5 leaps. The dog, however, covered as much space in 7 leaps as the rabbit did in 9 leaps. How many leaps would the rabbit make before the dog overtook him?

Solution. If we make x the required number of leaps, then $100+x=$ the total number of leaps. Since the rabbit makes 6 leaps while the dog makes 5, the dog will make $\frac{5}{6}$ leaps while the rabbit makes 1. When the rabbit has made x leaps the dog has made $\frac{5x}{6}$ leaps, in which time he has gone as *far* as the rabbit has in $100+x$ leaps.

Before we can place the two quantities equal to each other, we must reduce the dog-leaps to their equivalent in rabbit-leaps. Now since 7 dog-leaps are equivalent to 9 rabbit-leaps, 1 dog-leap is equivalent to $\frac{9}{7}$ rabbit-leaps, and $\frac{5x}{6}$ dog-leaps are equal to $\frac{9}{7}\cdot\frac{5}{6}x$ rabbit-leaps. We therefore have the equation:

$$\frac{9}{7}\cdot\frac{5}{6}x=100+x$$
$$15x=1402+14x$$
$$x=1400$$

138.

EXAMPLE 30. A dealer has 2 grades of wine. The better grade sells for $1.50 a bottle, the second quality for $87\frac{1}{2}$ cents. He desires to make an intermediate grade by mixing the two in such a proportion as will make its value $1.12\frac{1}{2}$ a bottle. How much of each kind must he take to make 200 bottles?

Solution. Let $x=$ the number of bottles required of the better quality, then $200-x$ will be the number of bottles of the cheaper grade. Each bottle of the x grade is worth 1.50, and each bottle of the $200-x$ grade is worth 0.875, hence the value of the mixture is $150x+87.5(200-x)$. But each bottle of the mixed wine is to be worth 112.5 hence the 200 bottles will be worth $112.5+200$ which gives the equation:

$$150x+87.5(200-x)=112.5\times200$$
$$300x\times175(200-x)=225\times200$$
$$300x+35000-175x=45000$$
$$125x=10000$$
$$x=80$$

Hence 80 bottles of the better wine, and $200-80=120$ bottles of the cheaper wine, will be required.

139.

EXAMPLE 31. The Fahrenheit thermometer has its scale divided from 0° to boiling, into 212 degrees. The Centigrade thermometer starts at a point $=32°$ on the Fahrenheit scale, and from this 0° to boiling is divided into 100 degrees. When the mercury stands at 77 degrees Fahrenheit, what will it be in degrees Centigrade?

Solution. Let $x=$ required number of degrees C. Since the 0° Centigrade scale starts at 32° Fahrenheit the same distance which is divided into 100 degrees Centigrade, is divided into $212-32=180$ degrees Fahrenheit, and hence 1 degree C is equal to $\frac{180}{100}$ degrees F, and x degrees C, is equal to $\frac{180}{100} x$ degrees C. But before we can make an equation we must find out how many Fahrenheit degrees 77 F. is above the Centigrade 0°, and as the Fahrenheit degrees start 32 degrees F below the 0 C, we must subtract 32. Hence we have

$$\tfrac{180}{100} x = 77 - 32$$
$$18\, x = 770 - 320$$
$$18\, x = 450$$
$$x = 25$$

hence, $\quad 77° \text{ F} = 25° \text{ C}$

140.

EXAMPLE 32. A bicycle which is geared to 70 inches (i. e., is geared so that one revolution of its cranks propels it as far as one revolution of a 70 inch wheel), has 8 sprockets on the rear wheel and 20 sprockets on the crank wheel. Suppose the crank sprocket is to be changed so that the gear will be reduced to 63; what will be the number of sprockets in the new sprocket wheel, the rear sprocket remaining unchanged?

Solution. Let the required number of sprockets be $=x$. The ratio of the sprockets in the first case is $\frac{20}{8}$, and in the second case $\frac{x}{8}$. But in the first case the wheel revolves 70 times while in the second it revolves 63 times, and the lower gear gives it the lesser number of revolutions. Hence we have the equation

$$\tfrac{x}{8} 70 = \tfrac{20}{8} 63$$
$$70x = 20 \cdot 63$$
$$7x = 126$$
$$x = 18$$

141.

EXAMPLE 33. A piece of lead weighing 23 pounds in the air, is found to weigh 21 pounds when immersed in water, so that we may say 23 pounds of lead loses 2 pounds in water, and that this *proportion* holds good for any weight of lead, i. e. 2 times 23 pounds would lose $2 \times 2 = 4$ pounds in water. In the same way we find that 37 pounds of tin loses 5 pounds when weighed in water. If now we melt 23 pounds of lead and 37 pounds of tin together, the alloy will weigh 60 pounds, and it will, when weighed in water, lose $2 + 5 = 7$ pounds. Knowing these facts, we have given a mass of alloy, composed only of lead and tin, which weighs 217 pounds, and is found to lose 26 pounds when weighed in water. How much tin does it contain?

Solution. Suppose that the 217 pounds of alloy contains x pounds of tin, and hence $217 - x$ pounds of lead. Now the alloy must have lost 5 pounds in water for every 37 pounds of tin it contained, or $\frac{x}{37} \cdot 5$, and also have lost 2 pounds for every 23 pounds of lead, $\frac{x}{37} \cdot 5 + \frac{217-x}{23} \cdot 2$, and the sum of these must equal the loss of 26 pounds. Hence we have:

$$\frac{x}{37} \cdot 5 + \frac{217-x}{23} \cdot 2 = 26$$

whence, $\qquad 115x + 74 \cdot 217 - 74x = 23 \cdot 26 \cdot 37$

$$41x = 6068$$
$$x = 148$$

Hence the composition contained 148 pounds tin and 69 pounds lead.

142.

EXAMPLE 34. King Hiero of Syracuse gave a goldsmith 16 pounds of gold and 4 pounds of silver to make for him a crown. The completed crown weighed correctly 20 pounds, but the king suspected that the goldsmith had kept part of the gold, replacing it by an equal weight of silver. He therefore requested the mathematician Archimedes to investigate the crown for him. Archimedes weighed the crown in water, and found it to lose $1\frac{1}{4}$ pounds. He also found that 21 pounds of pure silver lost 2 pounds in water, and 20 pounds of gold lost 1 pound in water, and from these data he determined the amount of gold in the crown. How much gold did it contain?

Solution. Let the amount of gold $=x$, then $20-x=$ the silver. Then $\frac{x}{20}\cdot 1+2\frac{20-x}{21}\cdot 2\cdot$ must be equal to the whole loss in water, or $=1\frac{1}{4}$ pounds. Hence we have :

$$\frac{x}{20}+2\frac{20-x}{21}=\frac{5}{4}$$
$$\text{whence,} \quad x=14\frac{9}{19}$$

Hence, instead of 16 pounds of gold, the crown contained only $14\frac{9}{19}$ pounds, the jeweler having stolen $1\frac{9}{19}$ pounds, and put in instead 4 pounds of silver, $5\frac{10}{19}$ pounds.

143.

EXAMPLE 35. A man is 58 years old and his son 18 years. In how many years will the father be twice as old as his son?

Solution. Let x be the number of years which must elapse until the age of the father is twice that of the son. The father's age at that time will then be $58+x$, and the son's age $18+x$; but by the terms of the problem twice the son's age will equal the father's age and the equation will be :

$$2\,(18+x)=58+x$$
$$36+2x=58+x$$
$$x=22$$

This is the correct solution, for $58+22=80$ and $18+22=40$.

144.

EXAMPLE 36. A man's age is 58, and his son is 18; at what age is the father $3\frac{1}{2}$ times as old as his son.

Solution. Let x equal the number of years which must elapse for the father to be $3\frac{1}{2}$ times the age of the son, and their respective ages will then be $58+x$ and $18+x$, and we have :

$$3\frac{1}{2}\,(18+x)=58+x$$
$$63+3\frac{1}{2}x=58+x$$
$$2\frac{1}{2}x=58-63$$
$$\tfrac{5}{2}x=-5$$
$$x=-2$$

This calculation gives x a minus sign, which while perfectly correct requires some explanation.

The processes of calculation are perfectly logical, although they may in some cases appear to conflict with the conditions of the problem. Such conflict, however, is only apparent. The real problem which we have solved is that of the conditions :

$$3\tfrac{1}{2}(18+x)=58+x$$

which simply means that x is equal to a number which, united to 58 and 18 respectively, will make the former $3\tfrac{1}{2}$ times the latter, regardless of whether these quantities are ages, pounds, miles or anything else. Now when a quantity comes out negative, the minus sign simply means that the actual magnitude found for x is to be *subtracted* instead of added to the given ages. It must therefore be remembered that in solving an equation we may determine not only the *value* of the unknown, but also its character, i. e., whether it is positive or negative. As already explained, to *add* a negative quantity is the same as to *subtract* a positive quantity, and hence if we subtract 2 from each of the ages we get 56 and 16, for the ages of father and son respectively, showing that the relation desired took place in the past instead of the future.

145.

EXAMPLE 37. A man is 58 years old, and his son 18 years. How many years have passed since the father was six times as old as his son?

Solution. Let x be the number of years to be subtracted to make the father's age 6 times that of the son, and we have :

$$58-x=6(18-x)$$
$$58-x=108-6x$$
$$5x=50$$
$$x=10$$

Hence the father's age was $58-10=48$, and the son's, $18-10=8$ years.

146.

EXAMPLE 38. Again, let the father be 58 and the son 18 years, how much must be subtracted from both ages for the father to be three times the son's age?

Solution. Let the number of years to be subtracted $=x$, then the father's age will be $58-x$, and the son's age $18-x$, and the equation will be :

$$58-x=3(18-x)$$
$$58-x=54-3x$$
$$2x=-4$$
$$x=-2$$

Here again we have a negative result, and as the problem was to find how much was to be *subtracted*, we must *subtract* minus 2, which is the same as adding plus 2, and the ages are 60 and 20 years respectively.

147.

EXAMPLE 39. What must be the value and *sign* of x, to satisfy the following equation?

$$25+4x=7-2x$$

Solution. We have : $6x=-18$

$$x=-3$$

Substituting this value $-3=x$ in the original equation, we get

$$25+4(-3)=7-2(-3)$$
$$25-12=7+6$$

This example shows, that the absolute sign of the unknown quantity can only be determined when it appears as an inverse factor, as will be explained hereafter, see § 329, appendix.

BOOK XIII.

Functions and Formulas.

148.

In mathematical works, we often meet with the statement that a certain quantity is a *function* of one or more other quantities. This expression means simply that there exists some definite *relation* between the quantities named, the one depending upon the other (or others) according to some definite laws.

Thus, for example, the distance to which a cannon ball may be fired depends upon several other quantities, such as the initial velocity, the size and weight of the ball, the composition and explosive force of the powder, the angle of elevation of the gun, the length of its bore, the resistance of the air, the attraction of gravity, &c., &c. And since the distance is definitely affected by all these conditions, it is said to be a *function* of their values, depending upon them for *its* value.

149.

In further explanation of the preceding paragraph the following simple equation is given.

What number is it, of which the sum of its 5th and 7th parts, is equal to 24.

Solution. It is clear that the value of the required number (x), by the conditions of the question, is dependent upon the numbers 5, 7 and 24, or in mathematical language, x is a *function* of 5, 7, and 24, the relation being :

$$\tfrac{x}{5} + \tfrac{x}{7} = 24$$
$$\text{whence} \quad 7x + 5x = 35 \cdot 24$$
$$12x = 840$$
$$x = 70$$

Here we have readily found the required number to be $x=70$, but the precise relation which it bears to the numbers 5, 7, and 24, cannot be seen when they are thus collected, but must be otherwise expressed in the shape of what is called a *formula*, which simply means that the relations which the functions bear to each other must be placed in a definite *form*. In order to do this we must determine the value of x in the equation, not by performing the operations arithmetically, but by indicating them, while keeping the various quantities intact so that their relations are exhibited at one time before the observer in a *formula*.

Thus we have from the equation :

$$\tfrac{x}{5}+\tfrac{x}{7}=24$$

first clearing of fractions by multiplying both sides by $5 \cdot 7$

$$7x+5x=5 \cdot 7 \cdot 24$$

then indicating the collection of the coefficients of x, we get

$$(7+5)\,x=5 \cdot 7 \cdot 24$$
$$\text{whence} \quad x=\frac{5 \cdot 7 \cdot 24}{7+5}$$

Here we have a formula, which shows at a glance the manner in which the value of x depends upon 5, 7. and 24, this then is the *formula* for the value of x, which expressed in words, is that to find x, we must multiply the three numbers together and divide the product by the sum of the first two.

150.

It is easy to see that the law which shows how one quantity depends upon others is not determined by the absolute magnitude of the quantities, but by the relations which they bear to each other. Thus if it is asked what number it is of which the sum of the 3rd and 4th parts equals 21, we might state the equation and find the value of x as in the preceding paragraph. Instead of repeating the work, however, we have only to substitute in the final formula of that paragraph 21 for 24, and 3 and 4 instead of 5 and 7, and obtain at once the required number $=36$.

We may also express the law or conditions by which one quantity depends upon others, and find the simplest expression of their relations without losing their identity, by representing all the quantities by determinate symbols. Then stating the question as an equation, we can reduce it to its simplest form to obtain the desired formula. (see § 82).

The result which is obtained by the reduction of a general expression will clearly be the same, when letters are substituted for the numerical quantities, except that the arithmetical operations will only be indicated instead of being performed, and by · means of the reductions the actual amount of reckoning will finally be much reduced.

Thus the preceding problems can be reduced to a general formula as follows :

Let it be required to find a number such that its mth and nth parts added together shall equal a.

Solution.   Let x be the unknown, and m, n, and a the given numbers.   The conditions of the problem stated as an equation are :

$$\frac{x}{m} + \frac{x}{n} = a$$

clearing of fractions by multiplying both sides by mn, we get :

$$nx + mx = amn$$

Now adding the coefficients of x we have :

$$(n+m)\,x = mna$$

and dividing by the coefficient of x we have :

$$x = \frac{mna}{m+n}$$

151.

Example.   Let it be required to divide the number 140 into two parts, so that they shall be in the proportion of $2 : 5$.

Solution.   The parts must be in the proportion of $2 : 5$, which is the same as $1 : \frac{5}{2}$ (see § 63), and if we call one part x the other will be $\frac{5}{2}x$, whence we get the equation :

$$x + \tfrac{5}{2}x = 140$$
$$2x + 5x = 280$$
$$7x = 280$$
$$x = 40 = \text{one part}$$
$$\tfrac{5}{2}x = 100 = \text{the other part.}$$

Example.   Divide any number a, into two parts which shall be in the relation of $m : n$.

Solution. The proportion $m : n$ is the same as $1 : \dfrac{n}{m}$, so that if x be one part $\dfrac{n}{m}x$ will be the other, whence:

$$x + \frac{n}{m}x = a$$

multiplying by m, we get

$$mx + nx = ma$$
$$(m+n)x = ma$$
$$x = \frac{ma}{m+n} = \text{one part,}$$

and the other part will be:

$$\frac{n}{m}x = \frac{n}{m} \cdot \frac{ma}{m+n} = \frac{na}{m+n}$$

In order to prove the correctness of this work, we remember that the values for x and $\dfrac{n}{m}x$, added together should equal a. This is the case for we have:

$$\frac{ma}{m+n} + \frac{na}{m+n} = \frac{a(m+n)}{m+n} = a$$

152.

EXAMPLE. Divide a given number, a, into three parts, such that their relations shall be in the proportion of $m : n : p$.

Solution. The proportions $m : n : p$ are the same as $1 : \dfrac{n}{m} : \dfrac{p}{m}$ and hence if x is the first part, the second part will be $\dfrac{n}{m}x$, and the third part will be $\dfrac{p}{m}x$, and since the sum of all three is equal to a, we have:

$$x + \frac{nx}{m} + \frac{px}{m} = a$$

In order to reduce this to a value for x, we first multiply the whole equation by m, which gives:

$$mx + nx + px = ma$$

Then collecting the coefficients of x,

$$(m+n+p)x = ma$$

whence, $x = \dfrac{ma}{m+n+p}$

which is the first part, and the second part$=$

$$\frac{n}{m}x=\frac{n}{m}\cdot\frac{ma}{m+n+p}=\frac{na}{m+n+p}$$

and the third part $=$

$$\frac{p}{m}x=\frac{p}{m}\cdot\frac{ma}{m+n+p}=\frac{pa}{m+n+p}$$

Thus, for example, suppose it is required to separate 100 into three parts which shall be in the relation of $2:5:3$, we have in the above formulas: $a=100$, $m=2$, $n=5$, $p=3$, and the

$$\text{first part}=\frac{2\cdot100}{2+5+3}=\frac{2\cdot100}{10}=20$$

$$\text{the second part}=\frac{5\cdot100}{10}=50$$

$$\text{the third part}=\frac{3\cdot100}{10}=30$$

153.

EXAMPLE. Divide the number 100 into two parts, such that when one part is divided by $m=5$, and the other part by $n=8$, the sum of the quotients shall equal 17.

Solution. Let x be one part and $a-x$ the other part, and we have from the problem:

$$\frac{x}{m}+\frac{a-x}{n}=b$$

Multiplying both sides by mn, we get:

$$nx+ma-mx=mnb$$

Now separating the known from the unknown quantities:

$$nx-mx=mnb-ma$$
$$(n-m)x=m(nb-a)$$
$$x=\frac{m(nb-a)}{n-m}$$

Subtracting this from a, to get the value for $a-x$, we have:

$$a-x=a-\frac{m(nb-a)}{n-m}$$

This expression for $a-x$ can be further reduced as follows:

$$a-x=\frac{a(n-m)-m(nb-a)}{n-m}$$

Removing the parenthesis :

$$a-x=\frac{an-am-mnb+am}{n-m}$$

$$a-x=\frac{an-mnb}{n-m}$$

$$a-x=\frac{n(a-mb)}{n-m}$$

In order to prove the above work to be correct we can add the expressions for x and $a-x$ together, and see if they equal a.

This proves to be the case since we have :

$$\frac{m(nb-a)}{n-m}+\frac{n(a-mb}{n-m}=$$

$$=\frac{mnb-ma+na-nmb}{n-m}=$$

$$=\frac{na-ma}{n-m}=\frac{(n-m)a}{n+m}=a.$$

154.

EXAMPLE. Suppose a number of payments are to be made at the termination of various periods of time ; i. e. a certain sum s, at the expiration of m months, another sum s' at the end of m' months, a third sum s'' at the end of m'' months, &c.

The creditor desires to receive the entire amount of all the payments $=s+s'+s''$—&c, at one time. At what time should this payment be made, in order that it should be the correct equivalent of the separate payments at different times ? (It is often customary to indicate quantities of the same kind by the same letter, using accent marks to distinguish the various items. Thus above the letter s is used to represent a sum of money, and the different sums by s, s', s'', &c., while m, m', m'', &c., represent the various numbers of months.)

Solution. In order to be able to state the question clearly, let us choose any monthly percentage, and say that the interest per month, i. e. the number of dollars returned by each \$100 per month, shall $=p$. The total interest on a sum s for m months will then be $=\frac{spm}{100}$; and likewise the interest on a sum

s' will equal $\dfrac{s'pm'}{100}$; and so on. The total amount which would then be paid the creditor would then be

$$\frac{spm}{100}+\frac{s'pm'}{100}+\frac{s''pm''}{100}\ldots$$

the amount of the interest in each case depending upon the length of time. Now the time at which the whole sum $s+s'+s''+\ldots$ should be paid, to be of the equivalent value, we will call x, and this will also be the same time at which the interest on the total sum at the same rate p, would equal the sum of the interests on the different sums for the different times. We can therefore make the following equation :

$$\frac{(s+s'+s''..)px}{100}=\frac{spm}{100}+\frac{s'pm'}{100}+\frac{s''pm''}{100}+..$$

We see that all members have the factor $\dfrac{p}{100}$, which is evident also from the fact that the equality is true, whatever the rate of interest may be. We may then cancel this factor by multiplying both sides by $\dfrac{p}{100}$, and then by dividing both sides by $s+s'+s''+..$ we get :

$$x=\frac{sm+s'm'+s''m''+..}{s+s'+s''+..}$$

This may be also expressed in the following rule : *Multiply each amount by its time, and divide the sum of the products by the sum of the amounts. The quotient will be the average time for the whole amount.*

EXAMPLE. The four following amounts are to be paid at the expiration of the given times, without interest, viz.: $300 at 14 days, $200 at 3 weeks, $150 at 2 months, and $100 at 28 days. What will be the average time if the total sum is to be all paid at once?

Here we have :

$s=300$	$s'=200$	$s''=150$	$s'''=100$
$m=14$	$m'=21$	$m''=60$	$m'''=28$
$sm=4200$	$s'm'=4200$	$s''m''=9000$	$s'''m'''=2800$

$$sm+s'm'+s''m''+s'''m'''=20200$$
$$s+s'+s''+s'''=750$$
$$x=\tfrac{20200}{750}=27 \text{ days, very nearly.}$$

155.

In all cases in which a quantity is only an *apparent* function of another, as x and p, in the preceding example, the latter must disappear in the course of the reduction. For example, in the expression :

$$\left(\left[a - \frac{m(bn-a)}{n-m} \right] \frac{n-m}{n} + mb \right) \frac{x}{3a}$$

x appears to be a function of a, b, m and n; that is, its value appears to depend upon the values of these quantities. As a matter of fact, however, such is not the case. If we make $x=5$ the value of the expression will reduce to $1\frac{2}{3}$, no matter what values we give to a, b, m and n. This follows from the fact that the expression will reduce to $\frac{x}{3}$ if we remove the parentheses.

In the same way we find that the value of the expression:

$$\tfrac{1}{2}(a+b)+\tfrac{1}{2}(a-b)$$

is entirely independent of the value of b.

156.

When one quantity is a function of several other quantities, each one of the latter is also inversely a function of all the others. For example, the amount of interest is a function of the principal, of the rate per cent., and of the time. Inversely, the time in which a given amount of interest has accrued is a function of the amount of interest, of the principal and of the rate. If therefore, we can obtain an equation expressing the relation between any set of functions, expressed in symbols, we can reduce this equation to expressions or formulas for the value of each function in terms of the others, and thus obtain a very clear idea of the relations existing between them.

157.

Example. If the interest upon a given capital C, at a rate p, for n years, be added to the principal to form a new capital C', each one of these four quantities n, p, C, and C', is a function of the other three. The value of each in terms of the other is to be expressed in formulas.

Solution. The expression $\dfrac{Cpn}{100}$ is equal to the amount of interest for n years, and if we add this to the original capital C, we have for the new capital:

$$C' = C + \frac{Cpn}{100}$$

In order to obtain from this equation values for C, p, and n, we first multiply both sides by 100, which gives

$$100\, C + Cpn = 100\, C'$$

collecting the coefficients of C,

$$(100 + pn)C = 100C'$$

dividing by the coefficient of C,

$$C = \frac{100C'}{100 + pn}$$

which gives a formula for the value of C.

Again, from the equation,

$$100C + Cpn = 100C'$$

we have $\qquad Cpn = 100C' - 100C$

which divided by Cn gives

$$p = \frac{100(C' - C)}{Cn}$$

or divided by Cp gives

$$n = \frac{100(C' - C)}{Cp}$$

The required formulas therefore are:

$$C' = C + \frac{Cpn}{100} \qquad (1)$$

$$C = \frac{100C'}{100 + pn} \qquad (2)$$

$$p = \frac{100(C' + C)}{Cn} \qquad (3)$$

$$n = \frac{100(C' + C)}{Cp} \qquad (4)$$

EXAMPLE. What must be the principal which when added to its interest at $4\frac{1}{2}$ % for 6 years will give a capital of \$762?

Substituting in formula 2, the values, $n=6$, $p=4\frac{1}{2}$, $C'=762$ we get:

$$C = \frac{100 \cdot 762}{100 + 6 \cdot 4\frac{1}{2}} = \frac{76200}{127}$$

$$C = \$600$$

How long must a principal of $600 remain at 5%, for the principal and interest to amount to $800 ?

Here we have given $C=600$; $p=5$, $C'=800$, and n is required. From formula 4, we have :

$$n=\frac{100\,(800-600)}{5\cdot600}$$

whence $n=6\frac{2}{3}$ years.

Let us take another case. The distance which a bicycle travels for each revolution of the crank, is a function of the number of teeth in each sprocket wheel, and also of the circumference of the bicycle wheel. If we call the distance $=d$, the circumference of the wheel $=c$, the number of teeth in the large sprocket $=N$ and the number of teeth in the small sprocket $=n$, we can deduce formulas for the relations of these functions.

The *more* teeth the large sprocket wheel has the faster the small sprocket wheel will revolve for one revolution of the crank, because each tooth of the large sprocket drives one tooth of the small one. For the same reason, the *fewer* teeth the small sprocket wheel has, the faster it will revolve, under the same conditions. Hence the number of revolutions which the small sprocket will make for one revolution of the crank, will be $\frac{N}{n}$; i. e. if the large wheel has twice as many teeth the small wheel will make twice as many revolutions, &c. Now if we call r the number of revolutions which the small sprocket and its attached wheel make, for one revolution of the crank, we have :

$$r=\frac{N}{n}$$

The bicycle travels a distance equal to the circumference C, of the wheel, for every revolution, hence for r revolutions the distance will be

$$d=Cr$$

whence we get

$$r=\frac{d}{C}$$

Placing these two values of r, equal to each other we have :

$$\frac{d}{C}=\frac{N}{n}$$

or $\quad dn=CN$

whence $\quad d=\dfrac{CN}{n}$ $\qquad$ (1)

$$n=\frac{CN}{d} \qquad (2)$$

$$C=\frac{dn}{N} \qquad (3)$$

$$N=\frac{dn}{C} \qquad (4)$$

EXAMPLES. In a bicycle, the circumference C, of the wheel is $7\frac{1}{3}$ feet, and the teeth in the sprocket wheels are $N=21$, $n=9$. How far will the machine travel for each revolution of the crank? Substituting these values in formula (1) we have :

$$d=\frac{7\frac{1}{3}\cdot 21}{9}=17\frac{1}{9}\text{ft}.$$

Suppose we wish it to travel 22 feet for each revolution, and have 8 teeth on the small sprocket, and the same diameter wheel; how many teeth must be on the large sprocket?

Here we have $C=7\frac{1}{3}$, $n=8$, $d=22$, and these values in (4), give :

$$N=\frac{22\cdot 8}{7\frac{1}{3}}$$
$$=\frac{22\cdot 8}{\frac{22}{3}}=\frac{22\cdot 8}{\frac{22}{3}}$$
$$=\frac{22\cdot 8\cdot 3}{22}=24 \text{ teeth}.$$

158.

EXAMPLES. Reduce the following examples to give the expressions for the value of x.

(1) $\quad ax=b+\dfrac{cx}{m}$

(2) $\quad \dfrac{ax}{m}+\dfrac{bx}{n}-\dfrac{c}{a}=0$

(3) $\quad \dfrac{ax}{m}-1-\dfrac{bx}{n}+c=0$

(4) $\quad \dfrac{a}{x}+\dfrac{b}{x}=\dfrac{1}{x}+c$

(5) $\quad c=a+\dfrac{a(a-x)}{a+x}$

(6) $\dfrac{mx-(c-x)n}{m(2x-c}=1$

(7) $\dfrac{a(dd+x)}{dx}=ac+\dfrac{a}{d}$

(8) $\dfrac{c}{a+bx}=\dfrac{h}{d+ex}$

ANSWERS.

(1) $x=\dfrac{bm}{am-c}$;   (2) $x=\dfrac{mnc}{a(an+bm)}$;

(3) $x=\dfrac{mn(1-c)}{an-bm}=\dfrac{mn(c-1)}{bm-an}$;

(4) $x=\dfrac{a+b+1}{c}$ (5) $x=\dfrac{a(2a-c)}{c}$;

(6) $x=c$;   (7) $x=\dfrac{d}{c}$;

(8) $x=\dfrac{cd-ah}{bh-ce}=\dfrac{ah-cd}{ce-bh}$.

BOOK XIV.

Equations of the First Degree

With Several Unknown Quantities.

159.

In nearly every case in applied mathematics and in mechanical and physical researches it is found that the investigations lead to expressions involving the relations of several unknown quantities and that from the relations of these to other known quantities several different equations can be deduced, in which the solution of the problem must *satisfy all the equations at 'the same time.*

For example, in the three following equations, the values of x, y, and z, when the equations are correctly solved, will satisfy the conditions of all three equations alike.

$$\begin{array}{ll} x+y+z=14 & (1) \\ 2x+5y-4z=1 & (2) \\ 7x-2y+3z=25 & (3) \end{array} \Big\}$$

These three equations, expressed in words, amount to the following : There are three numbers so related to each other that, (1) the sum of all three $=14$; (2) twice the first, plus five times the second, minus four times the third $=1$, and (3) that seven times the first, minus twice the second, plus three times the third shall $=25$.

These conditions, when fulfilled, should give three numbers. which when substituted for x, y, and z, in the above equations will cause them to be true, or in other words will " satisfy " them. In the above case the values $x=5, y=3, z=6$, will satisfy the first and second equations, but will not satisfy the third.

To attempt to find by trial the true values for x, y, and z, which will fulfill all the conditions, would not only be most tedious and difficult, but also in many cases impossible, and hence we consider whether there may not be some general method by which several combined equations, with a number of unknown quantities may be directly solved.

In fact there are a number of methods, of which we here give three, as most generally useful. The student will do well to try and think out additional methods for himself.

160.

First method.   Elimination by substitution.

Let the given equations be :

$$x+y+z=14 \tag{1}$$
$$2x+5y-4z=1 \tag{2}$$
$$7x-2y+3z=25 \tag{3}$$

Now since any one of the unknown quantities (such as x) has the same value in all three equations, we could find its value from any of them if the values of the other two were known. If then we proceed with whichever of the three equations is the most convenient, and proceed to solve it just as if y and z were known quantities, we shall obtain an expression which will give the value of x in terms of the other two unknown quantities. If then we "substitute" this expression in the place of x, wherever it occurs in the other two equations, we shall obtain two new equations, containing only two unknown quantities, y and z; x being, as it is called, "eliminated."

We can then take one of these new equations, and from it obtain a value for a second unknown, (such as y) in terms of the third unknown, and substituting this in the other equation obtain a new equation containing only z. This we can solve, and then by substituting its value back in the preceding equations, find the other two unknown quantities.

Applying this method to the above equations we have from (1) :

$$x=14-y-z \tag{1'}$$

Now if we place the value for x in equations (2) and (3) we shall have eliminated x, thus :

$$2(14-y-z)+5y-4z=1$$
$$7(14-y-z)-2y+3z=25$$

which we then proceed to reduce, by removing the parenthesis and collecting the values of y and z. Thus the first becomes :

$$3y-6z=27$$
$$\text{or} \quad y-2z=-9 \qquad (4)$$

and the second becomes : $\quad 9y+4z=73 \qquad (5)$

Then from equation (4) we have :

$$y=2z-9 \qquad (4')$$

Substituting this value for y in equation (5) we have :

$$9(2z-9)+4z=73 \qquad (6)$$

which thus contains only z, and we find :

$$22z=154$$
$$\text{and} \quad z=7$$

Then substituting this value of z in (4') we get :

$$y=2\cdot7-9=5$$

and substituting these two values for y and z in (1') we have :

$$x=14-5-7=2$$
$$\text{Hence} \quad x=2, y=5, z=7$$

which values satisfy all three equations. .

161.

It is apparent that this method can be applied to the solution of any number of equations with the same number of unknown quantities.

If, for instance, we have six equations with six unknown quantities, x, y, z, t, v, w, we can reduce any one of the equations to a value for x, and substitute this value in the five remaining equations ; then reduce one of these five equations to a value for y, and substitute this value for y in the remaining four, &c., until the last equation is reduced and solved for the single remaining unknown quantity, and the others determined by reverse substitution as before. The work is sometimes simplified if all the equations do not contain all the unknown quantities. In such cases it will be found best to eliminate first those quantities which appear in the fewest equations.

For example, suppose we have the four equations :

$$3x-2v=13 \qquad (1)$$
$$2y+3v=1 \qquad (2)$$
$$5y-7z=25 \qquad (3)$$
$$3y-6z+5v=0 \qquad (4)$$

Solving the first equation for x, we have :

$$x=\frac{13+2v}{3} \qquad (1')$$

Now on looking over the other equations we see that we cannot substitute this value of x, since x does not appear in any of them. We see, however, that z appears in the 3rd and 4th equations and solving (3) for z, we get :

$$z=\frac{5y-25}{7} \qquad (3')$$

Substituting this in equation (4) we have :

$$3y-6\frac{(5y-25)}{7}+5v=0$$

and removing the parenthesis, and clearing of the fraction, we get :

$$21y-30y+150+35v=0$$
$$-9y+35v=-150 \qquad (5)$$

Next we take equation (2), and solving it for y, we have :

$$y=\frac{1-3v}{2} \qquad (2')$$

and substituting this value of y in (5) we obtain

$$-9\left(\frac{1-3v}{2}\right)+35v=150 \qquad (6)$$

whence $\quad -9+27v+70v=-300$
$$97v=-291$$
$$v=-3$$

Substituting this value in (2') we have :

$$y=\frac{1-3(-3)}{2}=\frac{1+9}{2}=5$$

This value of y, in (3') gives :

$$z=\frac{5\cdot5-25}{7}=0$$

and the value $v=-3$, substituted in (1') gives :

$$x=\frac{13+2\ -3}{3}=2\tfrac{1}{3}$$

Hence the four quantities are :

$$x=2\tfrac{1}{8}\ ;\ y=5\ ;\ z=0\ ;\ v=-3.$$

162.

Second Method. Elimination by reduction. Suppose we have again the equations :

$$x+y+3=14 \tag{1}$$
$$2x+5y-4z=1 \tag{2}$$
$$7x-2y+3z=25 \tag{3}$$

We solve all three of them for values of x, and obtain :

$$x=14-y--z \tag{1'}$$

$$x=\frac{1-5y-4z}{2} \tag{2'}$$

$$x=\frac{25+2y-3z}{7} \tag{3'}$$

Now since x has the same value in all, the three equations $(1')$, $(2')$, $(3')$ must all be equal to each other.

We can therefore place any two of them equal to each other and thus obtain equations free from x. Thus :

$$14-y-z=\frac{1-5y+4z}{2} \tag{4}$$

$$14-y-z=\frac{25+2y-3z}{7} \tag{5}$$

These new equations are now both solved for another unknown quantity, such as y, giving :

$$y=-9+2z \tag{4'}$$

$$y=\frac{73-4z}{9} \tag{5'}$$

These are also equal to each other and therefore we have :

$$-9+2z=\frac{73-4z}{9} \tag{6}$$

from which we get $z=7$, and by reverse substitution in $(4')$ and $(1')$ we get $y=5$, and $x=2$.

This method is also adapted for use with a greater number of equations and unknown quantities.

For example, we may determine the values of x, y, z and v, from the following equations, thus :

$$3x-2v=13$$
$$2y+3v=1$$
$$5y-7z=25$$
$$5v+3y-6z=0$$

$$x=\frac{13+2v}{3} \qquad (1)$$

$$y=\frac{1-3v}{2} \qquad (2)$$

$$z=\frac{5y-25}{7} \qquad (3)$$

$$z=\frac{3y+5v}{6} \qquad (4)$$

Equating (i. e. placing equal to each other) the two values of z, we have

$$\frac{5y-25}{7}=\frac{3y+5v}{6}$$

whence

$$y=\frac{150+35v}{9} \qquad (5)$$

then equating the values of y in (5) and (2),

$$\frac{150+35v}{9}=\frac{1-3v}{2}$$

and $v=-3$.

This value in (1), and (2), gives $x=2\frac{1}{3}$ and $y=5$, and the value $y=5$, substituted in (3), gives $z=0$.

163.

Third Method. Elimination by Addition or Subtraction.

This method is to be preferred to either of the others, when the coefficients of the same unknown quantity, are alike in two equations, or when they can readily be made so. It is generally preferable also, when the coefficients are letters.

Since the values of the unknown quantities are not changed when all the members on both sides of an equation are multiplied or divided by the same number, it is usually easy to make the coefficients of the same unknown quantity alike in two equations. This being done, we have only to subtract one equation from the other if the signs of the quantities are alike, or add them together if the signs are different, and we obtain a new equation with the unknown quantity eliminated. In the same manner we can proceed with two more equations, and so from n equations obtain $n-1$ equations and one less unknown quantity. By continuing in this way we proceed until we obtain but one unknown quantity, and then proceed as in the other methods.

Taking again the same equations

$$x+y+z=14 \qquad (1)$$
$$2x+5y-4z=1 \qquad (2)$$
$$7x-2y+3z=25 \qquad (3)$$

Multiplying equation (1) by 2, we have

$$2x+2y+2z=28$$

and subtracting equation (2) from this,

$$2x+2y+2z=28$$
$$2x+5y-4z=1$$

gives $\qquad -3y+6z=27 \qquad (4)$

Then again multiplying (1) by 7, and subtracting (3) we have

$$7x+7y+7z=98$$
$$7x-2y+3z=25$$
$$9y+4z=73 \qquad (5)$$

We now examine (4) and (5) and see that if we multiply (4) by 3 it will give y a coefficient of -9, and then adding the two equations together gives:

$$-9y+18z=81 \qquad (4')$$
$$9y+\ 4z=73 \qquad (5)$$
$$22z=154$$

Whence $z=7$, and this value in (5) or (4') gives $y=5$, and the values of y and z in (1) give $x=2$.

EXAMPLE. Find the values of x, y, z and v, from the following equations :

$$3x-2v=13 \qquad (1)$$
$$2y+3v=1 \qquad (2)$$
$$5y-7z=25 \qquad (3)$$
$$3y-6z+5v=0 \qquad (4)$$

The unknown quantity x appears in only one equation and hence cannot be eliminated. Since y and v occur in three equations, and z only in two, it is manifestly simpler to eliminate z first.

By multiplying the 3rd equation by 6, and the 4th by 7, we shall give z the same coefficient in both equations, since $6 \cdot 7 = 7 \cdot 6 = 42$, and as the signs are alike, we subtract. The operations are as follows:

$$30y - 42z \qquad = 150$$
$$21y - 42z + 35v = 0$$
$$\overline{\qquad 9y - \qquad 35v = 150} \qquad (5)$$

Multiplying (5) by 2, and (2) by 9, and subtracting, we get:

$$18y + 27v = 9$$
$$18y - 70v = 300$$
$$\overline{\qquad 97v = -291}$$
$$v = -3$$

This value substituted in the other equations gives the other unknown quantities as before:

$$x = 2\tfrac{1}{3}; \qquad y = 5; \qquad z = 0.$$

164.

EXAMPLE. Find the values of x, y, z and v, from the following set of equations:

$$x + y + z + v = 16 \qquad (1)$$
$$-x + y + 2z + 3v = 33 \qquad (2)$$
$$2x + 5y + 5z - 6v = 0 \qquad (3)$$
$$3x + 4y - z - 2v = -4 \qquad (4)$$

Solution. Adding (1) and (2) together gives

$$2y + 3z + 4v = 49 \qquad (5)$$

eliminating x. Then multiply equation (2) by 3 and add it to equation (3), giving:

$$7y + 9z = 66 \qquad (6)$$

Then multiply equation (2) by 3 and add it to equation (4), and we get:

$$7y + 5z + 7v = 95 \qquad (7)$$

We now have v only in two equations, and multiplying equation (5) by 7, and equation (7) by 4, and subtracting, we get:

$$-14y + z = -37 \qquad (8)$$

Then multiplying equation (6) by 2 and adding it to equation (8), gives:

$$19z=95 \qquad (7)$$

Whence $z=5$; from equation (6), $y=3$; from equation (5), $v=7$; and from equation (1) $x=1$.

165.

When an equation contains only one unknown quantity, we can find from it a determinate value for the unknown. When, as we have just shown, we have n equations with n unknown quantities (n being any number), we can reduce them to one equation with one unknown, which can then be determined, and thus by reverse substitution enable us to find determinate values for the n unknown quantities. To do this we must have as many *independent* and *non-contradictory* equations as there are unknown quantities.

For these reasons, problems which require the determination of any number of unknown quantities from the same number of independent equations, are called *determinate* problems.

In the application of the above methods of elimination it is necessary to take *all* the equations into account. If, for instance, in the above examples, we had combined the first equation with the second, then the first with the third, and not including the fourth equation at all, we should have obtained these equations freed from x, but on proceeding farther we would have obtained a so-called *identical* equation, i. e., one which would have reduced to the obvious but useless statement that the unknown was equal to itself, or that $0=0$. This error, which naturally follows from arguing in a circle, is one into which even expert mathematicians sometimes fall.

166.

When we have not as many independent equations as there are unknown quantities, the problem is said to be *indeterminate*, and this occurs for the reason that the conditions of the problem admit of an indeterminate number of solutions. In such cases there will be as many unknown quantities left undetermined as there are equations lacking.

Suppose for instance we have the very simple equation :

$$x+y=12 \qquad (1)$$

whence, $x=12-y$

We see at once that unless y be given we cannot determine x. Now if y cannot be determined from any other independent equation, the statement remains indeterminate. It tells us, however, the *relative* values of x and y. Thus, if $y=1$, x will $=11$; if $y=2$, $x=10$; if $y=3$, $x=9$, etc., etc., and any pair of these values will satisfy equation (1).

Again, if it be required to solve the following two equations:

$$4x-3y-4z=12 \qquad (1)$$
$$4x+y-8v=48 \qquad (2)$$

We get from equation (1) :

$$x=\frac{12+3y+4z}{4} \qquad (1')$$

Then subtracting equation (1) from (2) we have :

$$4y-8v+4z=36$$

whence, $y=9+2v-z$

This gives an equation in which two quantities are indeterminate, and for any values which may be given to them, corresponding value for x and y will be found. Thus, if $z=1$, $v=2$, y will $=12$ and $x=13$; for $z=1$, $v=1$, y will $=10$, and $x=11\frac{1}{2}$, etc. *Any* four such values will satisfy equations (1) and (2).

The theory of Indeterminate, or so-called Diophantine problems, which treats of the properties of numbers and other profound subjects, is one of very great scope, and whole treatises have been written upon the subject, among which are the *"Disquisitiones Arithmeticae,"* of Gauss ; the *"Theorie des Nombres,"* by Euler ; Fermat ; Legendre, and many others. We must here, however, limit ourselves to these few examples.

Another point must be noted in this subject. The several equations from which several unknowns are to be determined must be truly *independent*, and not such that one can be derived from the other merely by an arithmetical operation.

Thus the two equations :

$$x+y=12 \qquad (1)$$
$$2x+2y=24 \qquad (2)$$

are not independent, since the second is merely the first multiplied by 2, and the two equations really form but one.

In the same way the three following equations are not independent:

$$2x-3y=16 \qquad (1)$$
$$5x-12y-5z=47 \qquad (2)$$
$$3x+5z=17 \qquad (3)$$

The third equation is contained in the other two, as will readily be seen, if we multiply the first by 4 and subtract the second. We thus have really only two equations and as there are three unknown quantities the problem is indeterminate. This identity of equations is often concealed so that it does not appear until the latter part of the calculation, when the last equation will be found to contain more than one unknown quantity.

It is also necessary, if the problem is to be a determinate one, that none of the equations shall be contradictory. Thus in the two equations:

$$2x+3y=5 \qquad (1)$$
$$2x+3y=17 \qquad (2)$$

there is an evident contradiction, since the sum of two quantities cannot at one and the same time be equal both to 5 and to 17.

Finally, it sometimes occurs that there are many *more* equations resulting from the conditions, than there are unknown quantities. Such cases, although possessing much practical value and interest, involving the Theory of Probabilities, and the Method of Least Squares, belong to one of the most difficult portions of higher mathematics, and can only be referred to here.

167.

PROBLEM. Let it be required to find two numbers, of which the sum is 12 and the difference 6.

Solution. Let x be one number and y the other, and we have from the conditions of the problem, the following independent equations.

$$x+y=12 \qquad (1)$$
$$x-y=6 \qquad (2)$$

Adding (1) and (2) together we get:

$$2x=18$$
$$\text{whence,} \quad x=9$$

Subtracting (2) from (1), we have:

$$2y=6$$
$$\text{whence,} \quad y=3$$

The problem of finding two unknown quantities from their sum and difference, these latter being known, is one which often occurs. We find that *one of the numbers is equal to one half of the sum plus the difference, and the other is equal to one-half of the sum minus the difference.*

Thus if we call the sum *s*, and the difference *d*, we have:

$$x+y=s \qquad (1)$$
$$x-y=d \qquad (2)$$

Adding (1) and (2) we have:

$$2x=s+d$$

whence, $\qquad x=\dfrac{s+d}{2}$

Subtracting (2) from (1), we get:

$$2y=s-d$$

whence, $\qquad y=\dfrac{s-d}{2}$

168.

PROBLEM. There are two heaps of coins, such that if 10 pieces are transferred from the second to the first, the first will contain half as many pieces as the second. If, however, 30 pieces be transferred from the first to the second, the second will contain 6 times as many as the first. How many pieces are in each heap?

Solution. Let the number in the first heap be $=x$, and that in the second heap $=y$. If we take ten pieces from the second heap, it will contain $y-10$, and the first will contain $x+10$ pieces, and since the first then contains half as many as the second, we have:

$$x+10=\frac{y-10}{2} \qquad (1.)$$

If we take 30 pieces from the first and add them to the second, we have $x-30$ and $y+30$, and since then the second has 6 times the first we have:

$$y+30=6(x-30)$$

Transposing and collecting we get: $\qquad\qquad$ (2)

$$2x-y=-30 \qquad (1')$$
$$6x-y=210 \qquad (2')$$

Subtracting (1') from (2'), we have:

$$4x = 240, \quad x = 60$$

Multiplying (1') by 3 and then subtracting it from (2'), we have:

$$2y = 300, \quad y = 150$$

169.

PROBLEM. There is a fraction, such that if 1 be subtracted from both numerator and denominator, will equal $\frac{1}{5}$, and if 4 be added to both numerator and denominator, will equal $\frac{2}{5}$. What is the fraction?

Solution. If x be the numerator and y the denominator, the fraction will be represented by $\frac{x}{y}$, and we have from the conditions:

$$\frac{x-1}{y-1} = \frac{1}{5} \qquad (1)$$

$$\frac{x+4}{y+4} = \frac{2}{5} \qquad (2)$$

Multiplying (1) by $y-1$, and solving for x, we have:

$$x - 1 = \frac{y-1}{5} \qquad (1')$$

$$x = \frac{y-1}{5} + 1$$

Multiplying (2) by $y+4$, and solving for x,

$$x + 4 = \frac{2y+8}{5} \qquad (2')$$

$$x = \frac{2y+8}{5} - 4$$

Equating these two values for x:

$$\frac{2y+8}{5} - 4 = \frac{y-1}{5} + 1$$

$$\text{whence,} \quad y = 16$$

$$\text{and} \quad x = 4$$

and the required fraction is $\frac{x}{y} = \frac{4}{16}$

170.

PROBLEM. A reservoir is so arranged that water can run into it from three separate pipes, A, B and C. When A and B are both running, it is filled in 10 minutes; when B and C are running, it is filled in 20 minutes; and when A and C are both running, it is filled in 15 minutes. How long will it take for each pipe alone to fill it, and how long for all three running together?

Solution. Let the quantity of water held by the reservoir be equal to 1, (i. e., *one* thousand gallons, *one* ton, etc.) Since A and B can give this much in 10 minutes, they can give $\frac{1}{10}$ as much in one minute, and in the same time B and C give $\frac{1}{20}$, and A and C, $\frac{1}{15}$.

If then we call the time for each pipe to fill the reservoir alone to be x, y and z minutes, we have A alone in one minute will give $\frac{1}{x}$ part; B alone, $\frac{1}{y}$ part; C alone, $\frac{1}{z}$ part. Therefore A and B together will give in one minute $\frac{1}{x}+\frac{1}{y}$; B and C together. $\frac{1}{y}+\frac{1}{z}$, and A and C together, $\frac{1}{x}+\frac{1}{z}$. These give the equations :

$$\frac{1}{x}+\frac{1}{y}=\tfrac{1}{10} \tag{1}$$

$$\frac{1}{y}+\frac{1}{z}=\tfrac{1}{20} \tag{2}$$

$$\frac{1}{x}+\frac{1}{z}=\tfrac{1}{15} \tag{3}$$

In these equations all the quantities appear as *denominators*, with the same numerator, and in such cases it is often more convenient to free the equations at once from fractions by substituting for the fractions simple symbols, such as x', y', z', and then re-substituting the original values after the equations have been solved. In the present case, however, we will proceed as follows :

Subtracting equation (2) from (1) we get :

$$\frac{1}{x}-\frac{1}{z}=\tfrac{1}{20} \tag{4}$$

Adding (3) and (4) together :

$$2\cdot\frac{1}{x}=\tfrac{7}{60}$$

whence, $\quad\frac{1}{x}=\tfrac{7}{120}$

Substituting this value of $\frac{1}{x}$ in (1) and (4), we have :

$$\frac{1}{y}=\tfrac{5}{120}, \text{ and } \frac{1}{z}=\tfrac{1}{120}$$

whence, $x=17\tfrac{1}{7}$, $y=24$, $z=120$

When all three pipes are open the flow per minute would be :

$$\tfrac{7}{120}+\tfrac{5}{120}+\tfrac{1}{120}=\tfrac{13}{120}$$

Whence, the time required to fill the reservoir would be $1 : \tfrac{13}{120}=9\tfrac{3}{13}$ minutes.

The same result could have been obtained much more quickly by adding equations (1), (2) and (3) together, giving :

$$\frac{2}{x}+\frac{2}{y}+\frac{2}{z}=\tfrac{13}{60}$$

Then dividing by 2 :

$$\frac{1}{x}+\frac{1}{y}+\frac{1}{z}=\tfrac{13}{120} \qquad\qquad (4')$$

Subtracting (1) gives z, subtracting (2) gives x, etc.

171.

PROBLEM. A tank can be filled by the pipes A and B, in a minutes, by pipes B and C, in b minutes, by pipes A and C in c minutes.

It is desired to derive a general formula by means of which one can determine the time required to fill the tank by each pipe alone, and by all together.

Solution. Let x, y, and z be the times required for each pipe alone, then we have :

$$\frac{1}{x}+\frac{1}{y}=\frac{1}{a} \qquad\qquad (1)$$

$$\frac{1}{y}+\frac{1}{z}=\frac{1}{b} \qquad\qquad (2)$$

$$\frac{1}{x}+\frac{1}{z}=\frac{1}{c} \qquad\qquad (3)$$

Subtracting (2) from (1), gives :

$$\frac{1}{x}-\frac{1}{z}=\frac{1}{a}-\frac{1}{b} \qquad\qquad (4)$$

Adding (3) and (4) together gives :

$$2\cdot\frac{1}{x}=\frac{1}{a}-\frac{1}{b}+\frac{1}{c}=\frac{ab+bc-ac}{abc}$$

whence, $\dfrac{1}{x} = \dfrac{ab + bc - ac}{2abc}$

$\dfrac{1}{y} = \dfrac{ac + bc - ab}{2abc}$;

$\dfrac{1}{z} = \dfrac{ab + ac - bc}{2abc}$;

Hence we have :

$$x = \dfrac{2abc}{ab + bc - ac} \; ;$$

$$y = \dfrac{2abc}{ac + bc - ab} \; ;$$

$$z = \dfrac{2abc}{ab + ac - bc} \, .$$

The time required for all three pipes to fill the tank will be :

$$\dfrac{1}{\dfrac{1}{x} + \dfrac{1}{y} + \dfrac{1}{z}}$$

and substituting the values found above for $\dfrac{1}{x}$, $\dfrac{1}{y}$, and $\dfrac{1}{z}$, this reduces to :

$$\dfrac{2abc}{ab + ac + bc}$$

172.

PROBLEM. From the two equations given below, find the values for x and y, in terms of a, b, c, m, and n.

$$\dfrac{xy}{ax + by} = m \tag{1}$$

$$\dfrac{3xy}{cx - by} = n \tag{2}$$

Solution.   We will reduce both equations to the value of x, first clearing of fractions.   From equation (1) we have :

$$xy = amx + bmy$$
$$xy - amx = bmy$$
$$(y - am)x = bmy$$

whence, $x = \dfrac{bmy}{y - am}$ $\tag{1'}$

From equation (2) we have:

$$3xy = cnx - bny$$
$$3xy - cnx = -bny$$
$$(3y - cn)x = -bny$$
$$x = \frac{-bny}{3y - cn} \qquad (2')$$

Equating these two values of x:

$$\frac{bmy}{y - am} = \frac{-bny}{3y - cn} \qquad (3)$$

Dividing both sides by the common factor by:

$$\frac{m}{y - am} = \frac{-n}{3y - cn}$$

Then multiplying by the common denominator $(y - am)(3y - cn)$, we have:

$$3my - cmn = -ny + amn$$
$$3my + ny = amn + cmn$$
$$(3m + n)y = mn(a + c)$$
$$y = \frac{mn(a + c)}{3m + n}$$

Substituting this value for y in (1'), we get:

$$x = \frac{bm \cdot \dfrac{mn(a + c)}{3m + n}}{\dfrac{mn(a + c)}{3m + n} - am} =$$

$$x = \frac{\dfrac{bmmn(a + c)}{3m + n}}{\dfrac{mn(a + c) - (3m + n)am}{3m + n}} =$$

$$x = \frac{bmmn(a + c)}{mn(a + c) - (3m + n)am} =$$

$$x = \frac{bmn(a + c)}{an + cn - 3am - an} = \frac{bmn(a + c)}{cn - 3am}$$

Hence the required values are:

$$x = \frac{bmn(a + c)}{cn - 3am}$$

$$y = \frac{mn(a + c)}{3m + n}$$

Such equations can often be solved much more simply by using skill in observation and inspection. Thus if we inverted equations (1) and (2), we would have :

$$\frac{ax+by}{xy}=\frac{1}{m}$$

and $$\frac{cx-by}{3xy}=\frac{1}{n}$$

by division we have :

$$\frac{a}{y}+\frac{b}{x}=\frac{1}{m}$$

$$\frac{c}{3y}=\frac{b}{3x}=\frac{1}{n}$$

These may be separated into :

$$a\cdot\frac{1}{y}+b\cdot\frac{1}{x}=\frac{1}{m}\ ;\ c\cdot\frac{1}{y}-b\cdot\frac{1}{x}=\frac{3}{n}$$

Now let us put $\frac{1}{x}=u\ ;\ \frac{1}{y}=v$, and these become

$$av+bu=\frac{1}{m} \qquad\qquad \text{A}$$

and $$cv-bu=\frac{3}{n}$$

Adding these together gives :

$$(a+c)\,v=\frac{1}{m}+\frac{3}{n}$$

or, substituting the value of v, gives :

$$(a+c)\cdot\frac{1}{y}=\frac{1}{m}\cdot\frac{3}{n}$$

and multiplying by mny gives :

$$mn(a+c)=ny+3my$$

whence, $$y=\frac{mn(a+c)}{n+3m}$$

Since now v and y are known, we obtain u from equation (A) and this gives us x, since $u=\frac{1}{x}$.

173.

PROBLEM. Required to find, from the following equations, the values of x and y in terms of a, b, c, d, m, and n.

$$ax+by=c \qquad\qquad (1)$$
$$mx-ny=d \qquad\qquad (2)$$

Solution. Multiplying (1) by n, and (2) by b, we get:

$$anx + bny = cn \qquad\qquad (1')$$
$$bmx - bny = bd \qquad\qquad (2')$$

Then adding these together we have:

$$anx + bmx = cn + bd$$
$$\text{or} \quad (an + bm)x = cn + bd$$
$$\text{whence,} \quad x = \frac{cn + bd}{an + bm}$$

In the same way, multiplying (1) by m, and (2) by a, and subtracting, we get:

$$bmy + any = cm - ad$$
$$\text{whence,} \quad y = \frac{cm - ad}{an + bm}$$

BOOK XV.

Preliminary Ideas about Powers and Roots.

Extraction of Square and Cube Root.

174.

When a number is multiplied by itself several times, this operation is indicated by writing above and to the right a small figure denoting the number of times the multiplication is performed; i. e. the number of times the main number is taken as a factor.

Thus if the number 7 is taken five times as a factor we may write $7 \cdot 7 \cdot 7 \cdot 7 \cdot 7$ or more briefly 7^5. In the same way 3^3 means $3 \cdot 3 \cdot 3$, or 3 taken as a factor three times; and $a^4 = aaaa$. In general terms a^n means a taken as a factor n times.*

175.

Every such expression as 7^5, 3^3, a^4, a^n, &c., besides being the statement of the product of a number of equal factors, is also called a *Power* of the given quantity. The original number is called the Base or Root, while the small number at the upper right hand, which indicates how often the root is taken as a factor, is called the *Exponent* of the power.

In the expression 2^3; 2 is the *root;* 3 is the *exponent,* and the result of the actual multiplication $2 \cdot 2 \cdot 2 = 8$, is the *power.*

The student must be very careful not to confuse the exponent with the coefficient. For instance :

$$a^3 = a \cdot a \cdot a, \qquad \text{but}$$
$$3a = a + a + a.$$

---

*It is quite common to use the letter n (i. e. *number*) as a symbol when any *general* number of times is to be indicated.

Powers are really simplified expressions for the multiplication of like factors.

176.

Powers of different roots and exponents may have the same values. Thus $2^6=8^2=4^3=64$; $3^4=9^2=81$.

Powers are named by stating the *name* of the base (or root) and the *degree* of the exponent. Thus, 64 is the sixth power of 2, the second power of 8, and the third power of 4 ; likewise 81 is the fourth power of 3, or the second power of 9, &c., &c.

In general a^n is read a to the nth power, or more briefly a to the nth ; thus, 7^5 is read 7 fifth ; a^4 is read a fourth. The second power of a quantity is usually called the square, and the third power the cube. These terms are derived from their geometric significance. Thus, a^2 is read, a square, and is equal to aa ; a^3 is read a cube, and $=aaa$.

177.

The converse of a power is a *Root.* That is, when the power is given, the factor of which the power is composed, is called its Root, and the determination of this factor is called the "extraction" of the root. If $7^3=343$, (i. e. $7\cdot7\cdot7=343$) and we have given the power 343, and the number of times the factor is taken $=3$, and we wish to indicate the determination of the factor, we write :

$$\sqrt[3]{343}=7$$

The quantity 343 is then called the *base* of the root (or also the *power*), while the 3 is called exponent of the root, or frequently the *index* of the root, (sometimes the *degree* of the root). The expression $\sqrt[3]{343}$, is read the 3rd root of 343. In the same way $2^6=64$, and $\sqrt[6]{64}=2$; which is read, "the sixth root of $64=2$."

The symbol $\sqrt{}$ is called the "root" or "*radix*" symbol, sometimes the "radical" symbol ; (Latin "*radix*" meaning root), and the quantity under it, or affected by it, referred to as the "quantity under the 'radical.'" Sometimes for brevity an expression covered by the symbol $\sqrt{}$ is called as a whole "the radical," thus $\sqrt[3]{343}$ would be a "radical," and 7 would be the "value of the radical," or its "root."

It is evident that powers and roots are the opposite of each other, and hence can be used to prove each other. Thus $\sqrt[6]{64}=2$, because the root 2 with the exponent 6 is equal to 64, or $2^6=64$.

In the same way we see the $\sqrt[4]{81}=3$ is correct, because $3\cdot3\cdot3\cdot3=81$.

178.

In writing the second root the exponent, or index, is usually omitted, as being understood; so that we do not write $\sqrt[2]{}$, but simply $\sqrt{}$, and so whenever *no* index is given it is always understood to be the second root, or as it is always called, the "square" root. Instead of writing $\sqrt[2]{9}=3$; $\sqrt[2]{25}=6$, &c., we write $\sqrt{9}=3$; $\sqrt{25}=5$.

The third root is usually called the "cube" root. Thus, $\sqrt[3]{1000}=10$, which is read "the cube root of 1000 is $=10$.

179.

Every power of 1 is equal to 1. Thus $1^3=1\cdot1\cdot1=1$; and in general $1^n=1$, (read 1 to the nth power$=1$).

Conversely, every root of 1 is equal to 1. Thus $\sqrt[3]{1}=1$; $\sqrt[4]{1}=1$; and in general $\sqrt[n]{1}=1$.

180.

In order to raise a fraction to a given power, the numerator and denominator are both independently raised to the required power. Thus, for example:

$$(\tfrac{2}{3})^2=\frac{2^2}{3^2}=\tfrac{4}{9}; \text{ because } (\tfrac{2}{3})^2=\tfrac{2}{3}\cdot\tfrac{2}{3}.$$

$$(\tfrac{2}{3})^3=\frac{2^3}{3^2}=\tfrac{8}{27}; \text{ because } (\tfrac{2}{3})^3=\tfrac{2}{3}\cdot\tfrac{2}{3}\cdot\tfrac{2}{3}.$$

$$(\tfrac{1}{2})^3=\frac{1^3}{2_3}=\tfrac{1}{8}; \text{ for } (\tfrac{1}{2})^3=\tfrac{1}{2}\cdot\tfrac{1}{2}\cdot\tfrac{1}{2}.$$

Likewise $(\tfrac{4}{5})^2=\tfrac{16}{25}$; $(\tfrac{9}{4})^2=\tfrac{81}{16}$; $(\tfrac{1}{2})^2=\tfrac{1}{4}$; $(\tfrac{1}{3})^4=\tfrac{1}{81}$; $(1\tfrac{1}{4})^3=(\tfrac{5}{4})^3=\tfrac{125}{64}$.

In general :

$$\left(\frac{a}{b}\right)^n=\frac{a^n}{b^n}; \quad \left(\frac{1}{a}\right)^n=\frac{1}{a^n}.$$

Conversely: to extract the root of a fraction, we must extract the root of both numerator and denominator independently. Thus:

$$\sqrt{\tfrac{4}{9}}=\frac{\sqrt{4}}{\sqrt{9}}=\tfrac{2}{3};\qquad \sqrt[3]{\tfrac{8}{27}}=\frac{\sqrt[3]{8}}{\sqrt[3]{27}}=\tfrac{2}{3};$$

$$\sqrt[3]{\tfrac{1}{8}}=\tfrac{1}{2};\qquad\qquad \sqrt{\tfrac{16}{25}}=\tfrac{4}{5};$$

$$\sqrt{\tfrac{1}{4}}=\tfrac{1}{2};\qquad\qquad \sqrt{\tfrac{81}{16}}=\tfrac{9}{4};$$

$$\sqrt{\tfrac{9}{25}}=\tfrac{3}{5};\qquad\qquad \sqrt[n]{\tfrac{a}{b}}=\frac{\sqrt[n]{a}}{\sqrt[n]{b}}$$

181.

The higher the *power* to which a proper fraction is raised, the smaller the value becomes. Thus:

$$(\tfrac{2}{3})^{4}<(\tfrac{2}{3})^{2}$$

Conversely, however, the higher the degree of a *root* which is extracted from a proper fraction the greater the value becomes. Thus, we have:

$$\sqrt[4]{\tfrac{16}{81}}=\tfrac{2}{3}, \text{ and } \sqrt[2]{\tfrac{16}{81}}=\tfrac{4}{9}, \text{ and } \sqrt[4]{\tfrac{16}{81}}>\sqrt[2]{\tfrac{16}{81}}$$

Hence it is readily seen that for numbers *greater* than 1, the higher the power the greater the value, and the higher the root the smaller the value. We also see that since every root of 1 is equal to 1, that every root of a number greater than 1, must also be greater than 1; for instance $\sqrt[100]{2}>1$, or in general, $\sqrt[n]{2}>1$.

182.

When the numerator and denominator of a fraction are prime to each other, as $\tfrac{4}{5}$, or $\tfrac{2}{3}$, any power of it will also be a fraction of which the numerator and denominator are prime to each other, and is thus incapable of further reduction. (See § 319.)

No matter how often one multiplies a mixed number (such as $2\tfrac{1}{4}$, $3\tfrac{2}{3}$, $1\tfrac{2}{3}$, or their equivalent improper fractions, $\tfrac{9}{4}$, $\tfrac{11}{3}$, $\tfrac{5}{3}$, etc.) by itself, the resulting power will never become an unmixed whole number. Thus, for example:

$$(2\tfrac{1}{4})^{2}=(\tfrac{9}{4})^{2}=\tfrac{81}{16}=5\tfrac{1}{16};$$

$$(2\tfrac{1}{4})^{3}=(\tfrac{9}{4})^{3}=\tfrac{729}{64}=11\tfrac{25}{64};$$

$$(2\tfrac{1}{4})^{4}=(\tfrac{9}{4})^{4}=\tfrac{6561}{256}=25\tfrac{161}{256}.$$

See Appendix, §§ 316 and 319.

183.

When a root of a whole number is not an exact whole number, neither can it be an exact mixed number, nor can it have any *exact* or *complete* value. Thus, for example, the $\sqrt{4}=2$, $\sqrt{9}=3$, hence the square roots of all the numbers between 4 and 9 must lie between 2 and 3, and hence be fractional numbers. But, according to the preceding paragraph, no mixed number multiplied by itself can give an exact whole number, such as 5, 6, 7, 8. Hence the square roots of the numbers between these limits cannot be exact mixed numbers (that is, cannot be represented *exactly* by either proper or improper fractions). In like manner, since $\sqrt[3]{8}=2$ and $\sqrt[3]{27}=3$, the cube roots of all numbers between 8 and 27 must lie between 2 and 3, but cannot be exact mixed numbers, as none of these raised to the 3rd power can equal exact whole numbers.

184.

All roots between those which are exact whole numbers, such as $\sqrt{2}$, $\sqrt{3}$..., $\sqrt[3]{2}$, $\sqrt[3]{3}$..., $\sqrt[4]{2}$, $\sqrt[4]{3}$..., &c., are therefore called *irrational* or *incommensurable* quantities, because no *exact* numerical values can be found for them; while $\sqrt{1}$, $\sqrt{4}$, $\sqrt{\frac{4}{9}}$, $\sqrt{\frac{49}{9}}$, $\sqrt[3]{8}$, &c., are called *rational* or *commensurable* quantities.

Thus $\sqrt{\frac{81}{9}}$ is a rational quantity, since it can be *exactly* divided by $\frac{1}{3}$, 9 times (and is therefore commensurable). Since $\sqrt{\frac{81}{9}}=\frac{9}{3}$ we have the ratio $1 : \sqrt{\frac{81}{9}}$ as $1 : 9$, and every rational quantity must bear an exact ratio, in some proportion to unity, or else it would be incommensurable.

185.

Although the irrational quantities such as $\sqrt{2}$, $\sqrt[3]{2}$, $\sqrt[4]{2}$, &c., cannot be exactly measured by unity theoretically, yet by use of decimal fractions, they can be expressed to any desired degree of precision within the limits of human conception. Thus the following quantities; 2.4494898 and 2.4494897, differ from each other by one ten-millionth. Multiplying each one of these by itself we have

$$(2.4494898)^2=6.00000028$$
$$(2.4494897)^2=5.99999978$$

The first square differs by less than one-millionth, from 6, and the second is even closer to the exact value. Although therefore there is no number which when multiplied by itself will exactly equal 6, yet in practice we take 2.44949, or still more precisely 2.4494897 as the square root of 6. If more decimals are required, they can be obtained by the method hereafter explained in § 192.

We therefore have approximately:

$$\sqrt{6}=2.4495, \text{ or more exactly } \sqrt{6}=2.44948974.$$

The method of finding these decimals by the aid of logarithms; that is, finding the power of any root, or conversely, the root of any given power, will be explained in Book XX.

We must, however, here learn the detailed algebraic method of extracting the square and cube root. In Book XXI, we shall learn shorter and more convenient methods by using tables of logarithms, but the following complete methods should also be thoroughly mastered. For this purpose the following principles must be very clearly understood.

186.

Suppose a quantity composed of any two parts, which parts we may call in general a and b, so that the whole quantity is $a+b$. Now if this quantity is to be *squared*, or multiplied by itself, we have:

$$
\begin{array}{l}
a+b \\
a+b \\
\hline
a^2+ab \\
ab+b^2 \\
\hline
a^2+2ab+b^2
\end{array}
$$

We see that this square consists of three members, namely: *the square of the first part, twice the product of the first and second parts, and the square of the second part.*

This little formula, which we have here given both in symbols and in words, is most important, and should be *committed* to *memory* and always borne in mind. By its use the square of any quantity composed of two parts may be obtained without the labor of performing the multiplication, thus:

$$(x+a)^2=x^2+2ax+a^2$$

or, for example, $(76)^2=(70+6)^2=4900+840+36=5776$

187.

When a quantity composed of a number with zeros annexed is to be raised to any given power, the number is raised to the given power, and as many zeros annexed as there were in the original quantity, *multiplied by the exponent.* Thus, for example :

$$800^2 = 640000 \; ; \; 200^3 = 8000000 \; ; \; 1000^2 = 1000000.$$

188.

The square of any number of several figures, considered as a base, must have either twice as many figures as the base number, or twice as many, less one. Thus the square of a five figure number, as 33592, must consist either of $2 \cdot 5 = 10$ figures or $2 \cdot 5 - 1 = 9$ figures, since it must be something between the squares of the smallest five and six figure numbers. That is, it must fall somewhere between $(10000)^2 = 100000000$, (i. e., 9 figures) and $(100000)^2 = 10000000000$, (i. e., 11 figures), so that it cannot have fewer than 9 figures or as many as 11 figures.

The squares of all numbers composed of one figure, and the rational roots of all numbers composed of one and two figures, are contained in the multiplication table and are therefore assumed to be known.

189.

If we take any square number (i. e., any number of which the exact square root can be extracted) and starting at the right hand, point it off into portions of two figures each, the square root will have just as many figures in it as there have been portions pointed off in the original square number, Note that the last portion pointed off, that is the portion at the extreme left will have only one figure in it if the square number consists of an odd number of figures.

Also the first figure of the square root of the number will be the square root of the portion pointed off at the extreme left of the original number.

Thus in the number 80945678 we point it off thus 80'94'56'78 and see that there are four portions, hence there must be four figures in its square root. Also the greatest square in the left hand period is 64, and $\sqrt{64} = 8$, so the first figure of the square root will be 8.

Thus, we see that : .

$$81|00|00|00|=9000^2$$
$$80|94|56|78|=8xyz^2 \quad \text{(See § 187)}$$
$$64|00|00|00|=8000^2$$

and the square root of 80945678 must lie somewhere between 8000 and 9000, since 8000 is too small and 9000 is too large. In order now to find a method to determine the other figures x, y, and z we must first find the conditions by which this power is formed upon its base, or root, and then consider how the first and second figures, then the first two and the third, then the first three and the fourth, &c., are deduced, and so proceed from the first figure step by step to determine those which are unknown.

190.

Let us take any number composed of two, three, or four figures; such, for example, as 76, 764, or 7643, and in squaring it we can at once by inspection separate it into two parts thus :

$$76^2=(70+6)^2 \; ; \; 764^2=(760+4)^2$$
$$7643^2=(7640+3)^2.$$

If now we apply the formula

$$(a+b)^2=a^2+2ab+b^2,$$

to the first case, i. e. to $(70+6)^2$ we have $70=a$ and $6=b$, and the result will be :

$$
\begin{array}{r}
\overset{a}{7}0+\overset{b}{6})^2 \\
\hline
a^2=49|00| \\
2ab=\;\;8|40| \\
b^2=\;\;\;\;|36| \\
\hline
76^2=57|76|
\end{array}
$$

Having divided the power into two portions of two figures each (see § 189), we observe that the square of the first (or left hand figure $7^2=a^2$), lies entirely in the left hand portion of the power; the double product of the first and second figures, $(2ab=2(7+6)$ begins at the second figure of the first portion, while the square of the second figure $(b^2=6^2)$ lies entirely in the right hand portion of the power.

In the same way, we have the square of a three-figure number i. e., 764^2.

$$\overset{a}{(7\overset{\cdot}{6}0}+\overset{b}{4})^2$$

$$
\begin{array}{rr|r}
a^2= & 5776 & 00 \\
2ab= & 60 & 80 \\
b^2= & & 16 \\
\hline
764^2= 58 & 36 & 96
\end{array}
$$

Here the square of the first *two* figures of the root form the first two portions of the power; the double product lies in the second and third portions of the power; and the square of the last figure of the root lies in the last portion of the power. (Portions are always here meant the portions, of two figures each, pointed off from the right, as explained in § 189.)

For the square of a four-figure number such as 7643, we have:

$$\overset{a}{(7\overset{\cdot}{6}40}+\overset{b}{3})^2$$

$$
\begin{array}{rr|r}
a^2= & 58\ 36\ 96 & 00 \\
2ab= & 458 & 40 \\
b^2= & & 9 \\
\hline
7643^2= & 58|41|54|49 &
\end{array}
$$

In short, every "two-figure" portion of a power, corresponds to a single figure of the square root; and since any number of *portions* of the power cannot contain a larger square than the square of the corresponding number of *figures* of the root (counting from the left), we know that the greatest square root which is contained in the left hand period of the power, will be the first figure of the square root. Likewise the greatest square root which is contained in the first two periods of the power will be the first two figures of the root, etc.

191.

From the preceding paragraph we can frame the following rules for the extraction of the square root:

1. Separate the number, of which the square root is to be extracted, into portions or periods of two figures each, starting at the right and going toward the left; the extreme left hand period may contain one or two figures according as the number is odd or even.

Find by inspection the greatest square number (a^2) contained in the left hand period, and set its root down as the *first figure* (a)

of the required square root. Also write the square of this figure under the first period of the number and subtract it.

2. Bring down to the remainder the first figure of the next period of the number, and divide this by twice the root already found ($2a$) and place the quotient (b) as the second figure of the root. Also multiply it (b) by twice the first figure of the root ($2a$) and subtract the product ($2ab$).

3. Bring down to the remainder the second figure of the second period and subtract the square of the second figure of the root (b^2).

If there are still more periods to the number, the rules 2 and 3 are to be repeated, thus :

4. Bring down to the remainder the first figure of the next period ; divide by twice the *two* figures of the root already found, taking them both together now as the part (a) of the root. The quotient will be the third figure of the root. In order then to find another figure of the root, we always divide by twice the root already found. It may appear that the quotient comes out too large, so that its square cannot be subtracted, but this will immediately be seen and the next smaller taken.

The proof of the correctness of the work is the multiplication of the root by itself, as the product should be the original number again.

A few examples worked out and compared with the rules will make the latter quite clear. First a number of four figures :

$$
\begin{array}{r}
\overset{a\ b}{\sqrt{57|76}}=76 \\
a^2 \quad 49 \\
\hline
2a=14\ )\ 87 \\
2ab=\quad 84 \\
\hline
36 \\
b^2=\quad 36
\end{array}
$$

Here we point off 5776 into two parts, 57 | 76, and hence the root must consist of two figures. Then see by inspection that 49 is the greatest square number contained in 57, $36=6^2$ is too small, $64=8^2$ is too large, so $49=7^2$ is the nearest.

Subtracting 49 from 57, we have a remainder of 8. Bringing down the first figure of the next period, we get 87, and dividing this by twice 7, $=2a$, we see that it will go 6 times, that is

$2ab=84$. Subtracting, we have 3 remainder, and bringing down the remaining figure 6, we have 36, and this is just equal to $b^2=36$. We thus really have:

$$a^2+2ab+b^2$$
$$70^2+2(70 \cdot 6)+6^2$$
$$4900+840+36=5776$$

EXAMPLE 2.

$$\overset{a}{\overset{\overbrace{\quad}}{\underset{a\,b\,b}{}}}$$
$$\sqrt{58|36|96}=764$$

$$a^2=\quad 49$$

$$2a=14 \;) \;93$$
$$2ab=\quad 84$$

$$\quad 96$$
$$b^2=\quad 36$$

$$2a=152 \;) \;609$$
$$2ab=\quad 608$$

$$b^2=\quad 16$$
$$b^2=\quad 16$$

The above rules can easily be memorized either with or without the formula $a^2+2ab+b^2$. The work may also be greatly abridged, and the number of figures reduced to nearly one-half in the following manner. Instead of bringing down only the *first* figure of the second period, *all* the figures are brought down, the first one being separated from the rest by a comma, and the quotient multiplied directly by itself and by the divisor and both products subtracted at once, etc., as shown in the following examples:

(1)
$$\sqrt{57|76}=\overset{a\,b}{76}$$
$$a^2=\quad 49$$

$$2a,b=14,6 \;) \;87,6 \;(\;87:14=7$$
$$2ab+b^2=\quad 87 \;6 \;(=146 \cdot 6$$

(2)
$$\sqrt{58|36|96}=764$$
$$a^2=\quad 49$$

$$14,6 \;) \;93,6 \quad(93:14=6$$
$$87 \;6 \quad(=146 \cdot 6$$

$$152,4 \;) \;609,6 \;(609:152=4$$
$$609 \;6 \;(=1524 \cdot 4$$

(3) $\qquad \sqrt{58'41|54|49}=7643$

$\qquad\qquad\qquad 49$

$\qquad\ \ \ \overline{14,6\ )\ \ 94,1\qquad(\ 94:14=6)}$

$\qquad\qquad\quad 87\ 6\qquad\ \ (=146\cdot6)$

$\qquad\ \ \ \overline{152,4\ )\ 655,4\qquad(\ 655:152=4)}$

$\qquad\qquad\quad\ 609\ 6\qquad(=1524\cdot4)$

$\qquad\ \ \ \overline{1528,3\ )\ 4584,9\ (\ 4584:1528=3)}$

$\qquad\qquad\quad\ \ 4584\ 9\ (=15283\cdot3)$

(4) $\qquad\qquad \sqrt{4|03|20.64}=2008$

$\qquad\qquad\qquad 4$

$\qquad\ \ \overline{400,8\ )\qquad 32064}$

$\qquad\qquad\qquad\ \ 32064$

192.

In order to extract the root of a square whose root is irrational, as $\sqrt{6}$, for example, to any required number of decimal places, the following considerations must first be understood. We know that $6=\frac{600}{100}=\frac{60000}{10000}$, etc., etc., and from § 180, we have

$$\sqrt{6}=\sqrt{\tfrac{600}{100}}=\frac{\sqrt{600}}{\sqrt{100}}=\frac{\sqrt{600}}{10}=\tfrac{24}{10}=2.4$$

The square of 600, in round numbers is $=24$:

$$\sqrt{6\,00}=24$$
$$4$$
$$\overline{4,4)\ 20,0\ (20:4=4}$$
$$176=(44\cdot4$$
$$\overline{34}$$

Hence the $\sqrt{6}=2.4$ to the nearest tenth.   Also :

$$\sqrt{6}=\sqrt{\tfrac{60000}{10000}}=\frac{\sqrt{60000}}{\sqrt{10000}}=\frac{\sqrt{60000}}{100}$$

Now the square root of 60000 in round numbers $=244$, hence $\sqrt{6}$ is $=2.44$ to the nearest hundredth.

Hence we see how we can extract the square root of any incomplete square number, correct to as many decimal places as may be desired.

The root is first extracted in whole units, and the decimal point placed after the units of the root. Then annex two zeros to the remainder, treating these just as if they formed a period of the number, and thus proceed to find the tenth ; then annex two more zeros to the remainder, and find the hundredth, and so proceed until as many decimals have been found as are required.

Thus, the $\sqrt{283}=16.8226$ as follows :

$$\sqrt{2'83}=16.8226$$

$$1$$

2,6) 18,3 (18:2=6,7 is too large !)
15,6 (=26·6)

32,8) 270,0 (270:32=8)
262 4 (=328·8)

336,2) 760,0 (760:336=2)
627 4 (=3362·2)

3364,2) 87600 (8760:3364=2)
67284 (=33642·2)

33644,6) 203160,0
201867 6

1292400 &c.

Again we find $\sqrt{2}=1.41421\ldots$

$$\sqrt{2}=1.4142$$

$$1$$

2,4) 100 (10:2=4)
96 (=24·4)

28,1) 400 (40:28=1)
281 (=281·1)

282,4) 11900
11296

2828,2) 60400
56564

28284,1) 38360,0
28284 1

10075900 &c.

When a square root is required to several decimal places the work may be much abridged by the following method.

After the root has been extracted, say to three or four decimal places, it is only necessary to *divide the remainder* by *twice the root already found* in order to obtain two or three additional figures. This very convenient rule should always be remembered, and is based on the following principle:

If in the expression

$$(a+b)^2 = a^2 + 2ab + b^2,$$

b is a very small fraction, it is evident that b^2 will be still smaller.

Now if we call the part of the root already found $= a$, and the small fraction yet remaining to be found $= b$; we have the remainder, from our work already performed $= 2ab + b^2$, and this divided by $2a$ (that is twice the root already found), will give us $b + \dfrac{b^2}{2a}$ and since b^2 is very small, and is also made still smaller by being divided by $2a$, it may be neglected and the quotient of the remainder divided by $2a$, taken as equal to b. This will best be illustrated by application to one of the preceding examples.

$$\sqrt{2|83} = 16.822$$

$$
\begin{array}{r}
1 \\
\hline
2,6)\quad 183 \\
156 \\
\hline
32,8)\quad 2700 \\
2624 \\
\hline
336,2)\quad 7600 \\
6724 \\
\hline
3364,2)\quad 87600 \\
67284 \\
\hline
20316 \\
\end{array}
$$

Now if we wish additional decimals, we have :

$$16.822 = a$$
$$.000vxyz = b$$
$$.020316 = 2ab + b^2$$

And we have:

$$(16.822)^2 + .020316 + (.000vxyz)^2$$
$$a^2 \quad + \quad 2ab \quad + \quad b^2$$

Whence:

$$\frac{.020316}{2a} = b \text{ very nearly} = \frac{.020316}{33.644} = .0006038$$

and, $\sqrt{283} = 16.8226038$ correct to seven decimal places, of which the last four have been obtained merely by division.

It is evident that we can always obtain as many correct *additional* places as there are decimal places in the portion of the root already found. This is true because there will always be at least twice as many zeros in the square of any decimal as there are in the decimal itself. Thus if b were $= .009$, b^2 would $= .000081$ and hence the omission of b^2 from the root would not produce a significant effect before the fifth decimal place. Likewise if $b = .0009$, $b^2 = .00000081$ and as any smaller figure than 9 would produce a still smaller result, especially if divided by $2a$, a always being larger than b, we can always depend upon finding as many *additional* decimals correctly by this method as have already been determined by the previous work.

If we have a root to a certain number of decimals given, and desire additional decimals, but have not the previous work at hand, and so have not the remainder to work from, we proceed as follows: *Square the given root, and subtract the result from the given power. Divide this remainder by twice the given root, and the quotient will be the additional decimals, correct to as many more places as there were decimals originally given.*

Thus, suppose the square root of 774 is given as 27.820 nearly; we can find it to at least three more decimal places, as follows:

$$(27.820)^2 = 773.952400$$
$$774 - 773.952400 = 0.047600$$

$$\text{and,} \quad \frac{0.047600}{27.820 + 2} = .0008555$$

hence the more exact root is:

$$\sqrt{774} = 27.8208555,$$

and in this case the *seventh* decimal is correct.

193.

If we have a whole number with a decimal attached, we extract first the square root of the whole portion, and then instead of annexing two zeros, we bring down two figures from the decimal. In this way we find $\sqrt{312.506} = 17.677$

$$\sqrt{3|12|.50|60} = 17.677$$

$$
\begin{array}{l}
 1 \\
2,7)\ 212 \\
 189 \\
\hline
34,6)\ 235,0 \\
 207\ 6 \\
\hline
352,7)\ 2746,0 \\
 2468\ 9 \\
\hline
3534,7)\ 27710,0 \\
 24742\ 9 \\
\hline
 296710,0 \text{ etc.}
\end{array}
$$

In the same manner we find $\sqrt{0.00465} = 0.068$ as below :

$$\sqrt{0.00|46|50} = 0.06819$$

$$
\begin{array}{l}
\ 36 \\
\hline
12,8)\ 105,0 \\
 102\ 4 \\
\hline
136,1)\ 260,0 \\
 136\ 1 \\
\hline
1362,9)\ 123900 \\
 122661 \text{ etc.}
\end{array}
$$

We have in fact :

$$\sqrt{0.00465} = \sqrt{\tfrac{4650}{1000000}} = \sqrt{\tfrac{4650}{1000}} = \tfrac{68}{1000} = 0.068$$

194.

In extracting the square root of a common fraction it is most convenient to convert the fraction into a decimal, or else to multiply both numerator and denominator by the denominator, thus converting the denominator into a perfect square, and leaving only the root of the numerator to be extracted.

Thus, for example :

$$\sqrt{\tfrac{3}{7}}=\sqrt{0.428571}=0.654$$

$$\text{or} \quad \sqrt{\tfrac{3}{7}}=\sqrt{\tfrac{3}{7}\cdot\tfrac{7}{7}}=\sqrt{\tfrac{21}{49}}=\tfrac{4.582}{7}=0.654$$

In the same way :

$$\sqrt{2\tfrac{3}{8}}=\sqrt{2.3750}=1.541 \text{ or}$$

$$\sqrt{2\tfrac{3}{8}}=\sqrt{\tfrac{19}{8}}=\sqrt{\tfrac{38}{16}}=\frac{\sqrt{38}}{4}=\tfrac{6.164}{4}=1.541$$

The following examples should be worked out in full by the student :

$$\sqrt{76807696}=8764$$

$$\sqrt{1129969}=1063$$

$$\sqrt{3}=1.73205$$

$$\sqrt{5}=2.23606$$

$$\sqrt{8}=2.82842$$

$$\sqrt{10}=3.16227$$

$$\sqrt{25.0400057}=5.00399$$

$$\sqrt{25\tfrac{5}{7}}=5.07092$$

$$\sqrt{\tfrac{3}{5}}=0.77459$$

$$\sqrt{13\tfrac{5}{8}}=3.71483$$

$$\sqrt{\tfrac{3}{4}}=0.86602$$

$$\sqrt{\tfrac{7}{25}}=0.52915$$

$$\sqrt{0.0004}=0.02$$

195.

Extraction of the Cube Root.

In applying the customary arithmetical method to the extraction of roots of higher degree than the second, the practical difficulties increase very rapidly. Even the extraction of the cube root becomes such a tedious and fatiguing operation that even those who possess a great facility in numerical calculations make practical use of the shorter methods explained in the following chapters. There is, however, not the slightest difficulty in understanding the *theory* of the extraction of roots of higher degree, the method being quite similar to that of the extraction of the square root, already described. The following principles must first be explained.

196.

The cube of any number (or root) must have 3 times as many places of figures as the root, or 3 times as many less one, or 3 times as many less two.

The cube of a four place number, such as 4356, must therefore consist either of 12, 11, or 10 figures, since it must fall somewhere between the cubes of the next smaller and next larger simple numbers, i. e. between the cubes of 1,000 and 10,000; one of which has the same number of figures as the given root, and the other, one more figure. But according to § 187, the cube of 1,000 has 10 figures, and the cube of 10,000 has 13 figures.

The cube of any single place number, and conversely, the cube root of every perfect cube of 1, 2, or 3 place number, can be found from the following simple table, or reckoned mentally.

| Cube . . . . . . | 1 | 8 | 27 | 64 | 125 | 216 | 343 | 512 | 729 |
| Cube Root . . . | 1 | 2 | 3 | 4 | 5 | 6 | 7 | 8 | 9 |

This table is easily committed to memory and will be found most useful, especially in connection with the approximate method to be given hereafter.

197.

Now in order further to examine the construction of a cube root, we point off the number, of which the cube root is to be extracted, into periods of three places each, beginning at the right hand. Of course the extreme left hand period may contain one, two, or three figures, but the others will all consist of three figures each. Now having separated the number thus into periods, we find that the root must consist of as many places as there are *periods* in the original number, and also that the first figure of the root will be the cube root of the first (or left hand) period. Thus if the number be 635478923, for instance, we see that the root must consist of three figures, and that the first one will be 8, because we get three periods of three figures each, and 8 is the nearest cube root of the left hand period, thus:

$$512\ 000\ 000 = 800^3$$
$$635\ 478\ 923 = 8xy^3 \quad (\S\ 187)$$
$$729\ 000\ 000 - 900^3$$

The cube root of 635478923 must therefore lie between 800 and 900. In order now to derive a general rule by which we may determine the remaining figures x and y, we must examine the conditions which govern the relations of the various figures of the root to those of the cube.

198.

If we take a binomial quantity, $a+b$, and cube it, we obtain the following result:

$$
\begin{array}{l}
a+b \\
a+b \\
\hline
a^2+2ab+b^2 \\
\quad\quad a+b \\
\hline
a^3+2a^2b+\ ab^2 \\
\quad\ a^2b+2ab^2+b^3 \\
\hline
(a+b)^3 = a^3+3a^2b+3ab^2+b^3
\end{array}
$$

By keeping this formula in mind, we may apply it to the extraction of the cube root of any number, which for the purpose we divide into two parts, and in order to obviate the necessity for writing it down, we may frame it in words, as follows:

The cube of any number (which has been separated into two parts) is composed of the following four parts: *The cube of the first part, plus three times the product of the square of the first*

*multiplied by the second, plus three times the product of the first
multiplied by the square of the second, plus the cube of the second
part.*

In order to understand more readily the relations of the figures to each other in applying this rule we shall divide the root
into two such parts that the units shall form the second part and
the balance of the number with a zero annexed, shall form the
first part, thus for 74 we shall take 70+4, or for 748 we take
740+8. We then have:

$$(1) \qquad 74^3 = (\overset{a}{70} + \overset{b}{4})^3 = \begin{cases} 70^3 = 343\,000 = a^3 \\ 3 \cdot 70^2 \cdot 4 = 58\,800 = 3a^2b \\ 3 \cdot 70 \cdot 4^2 = 3\,360 = 3ab^2 \\ 4^3 = 64 = b^3 \end{cases}$$

$$74^3 = 405\,224$$

And in like manner:

$$(2) \qquad 748^3 = (\overset{a}{740} + \overset{b}{8})^3 = \begin{cases} 740^3 = 405\,224\,000 = a^3 \\ 3 \cdot 740^2 \cdot 8 = 13\,142\,400 = 3a^2b \\ 3 \cdot 740 \cdot 8^2 = 142\,080 = 3ab^2 \\ 8^3 = 512 = b^3 \end{cases}$$

$$748^3 = 418\,508\,992$$

And we can perceive from No. 1 the following facts: The
cube of the first figure of the root lies in the first period of the
number; the triple product of the square of the first by the second,
has one of its figures only in the second period, the triple product of the first by the square of the second, has two of its
figures in the second period; while the cube of the second part
lies entirely in the second period.

We also see by § 187 that when a cube number contains
more than two periods, the first group of two, three or more
periods cannot contain a greater cube than that of the corresponding two, three, or more figures of the root, and that the
same relations exist between the other portions of the power and
the root. We can therefore apply the formula $a^3 + 3a^2b + 3ab^2 + b^3$
to extract the cube root of any number, no matter into how
many periods it may be divided.

199.

Rule for the extraction of the cube root of any number:

1. Point off the number into periods of three figures each,

starting from the right hand end. The left hand (or first) period may contain 1, 2, or three figures.

2. Place for the first figure of the root the greatest root of which the cube is contained in the first period, and subtract its cube from the first period.

3. Bring down to the remainder the first figure of the next period; divide this by three times the square of the first figure of the root, and place the quotient as the second figure of the root. Multiply the quotient by 3 times the square of the first figure of the root, and subtract. (NOTE.—It is easy to make the mistake of taking the quotient too great, but this can readily be discovered by trial.)

4. To the remainder annex the second figure of the second period, and subtract three times the product of the first figure of the root by the square of the second figure.

5. To the remainder annex the remaining figures of the second period, and subtract the cube of the second figure of the root.

If the original number was composed of more than two periods, the foregoing rules, 3, 4 and 5, must be applied repeatedly, namely : annex to the remainder the first figure of the third period, divide by three times the square of the *first two* figures of the root already found, place the quotient as the third figure of the root, and proceed as in 3, 4 and 5.

It will be apparent that the numerical work will become much greater with each additional figure of the root.

The following examples will assist the student to understand the rules :

$$
\begin{array}{r}
\overset{a\,b}{\sqrt[3]{405|224}=74} \\
a^3=7^3= \quad 343\;:\;:\;: \\
3a^2=3\cdot7^2=147)\;\; 622\;:\;: \\
3a^2b=3\cdot7^2\cdot4=\quad\;\; 588\;:\;: \\
\hline
342\;: \\
3ab^2=3\cdot7\cdot4^2=\quad 336\;: \\
64 \\
b^3=64
\end{array}
$$

$$\sqrt[3]{418\,508\,992} = 748$$

$$
\begin{array}{rl}
a^3 = & 343 ::: ::: \\
\hline
3a^2 = 147) & 755 :: ::: \\
3a^2b = 3 \cdot 7^2 \cdot 4 = & 588 :: ::: \\
\hline
& 1670 : ::: \\
3ab^2 = 3 \cdot 7 \cdot 4^2 = & 336 : ::: \\
\hline
& 13348 ::: \\
b^3 = 4^3 = & 64 ::: \\
\hline
& 132849 :: \\
3a^2b = 3 \cdot 74^2 \cdot 8 = & 131424 :: \\
\hline
& 14259 : \\
3ab^2 = 3 \cdot 74 \cdot 8^2 = & 14208 : \\
\hline
& 512 \\
b^3 = 8^3 = & 512 \\
\hline
\end{array}
$$

The work is proved correct if the root when multiplied together three times equals the number from which it has been extracted.

200.

The foregoing rules can be applied to the extraction of irrational roots to any desired number of decimal places. After the whole units of the root have been found, periods of three zeros are successively annexed, and the tenths. hundredths, etc., found, as in square root (see § 192).

If the number consists of a whole number and a decimal, the method is the same, except three of the decimals are brought down instead of annexing three zeros. The process is the same for a number which is entirely composed of decimals. For a vulgar fraction it is most convenient either to convert it into a decimal fraction or to make the denominator a perfect cube by multiplying the whole of the fraction by the square of the denominator.

The following examples will illustrate these various operations :

$$\sqrt[3]{9\,007} = 20.806$$

$$
\begin{array}{rl}
a^3 = & 8 \\
\hline
3a^2 = 12) & 10 \\
\hline
3a^2 = 1200) & 10070 \\
3a^2b = & 9600 \\
\hline
& 4700 \\
3ab^2 = & 3840 \\
\hline
& 8600 \\
b^3 = & 512 \\
\hline
3a^2 = 129792) & 80880 \\
\hline
3a^2 = 12979200) & 80880000 \\
& \text{etc.}
\end{array}
$$

$$\sqrt[3]{2} = 1.25$$

$$
\begin{array}{rl}
a^3 = & 1 \\
\hline
3a^2 = 3) & 10 \\
3a^2b = & 6 \\
\hline
& 40 \\
3ab^2 = & 12 \\
\hline
& 280 \\
b^3 = & 8 \\
\hline
3a^2 = 432) & 2720 \\
3a^2b = & 2160 \\
\hline
& 5600 \\
3ab^2 = & 900 \\
\hline
& 47000 \\
b^3 = & 125 \\
& \text{etc.}
\end{array}
$$

$$\sqrt[3]{2.057\,600} = 1.2$$

$$
\begin{array}{rl}
& 1 \\
\hline
3a^2 = 3) & 10 \\
3a^2b = & 6 \\
\hline
& 45 \\
3ab^2 = & 12 \\
\hline
& 337 \\
& \text{etc.}
\end{array}
$$

$$\sqrt[3]{0.007|040}=0.19$$

$$
\begin{array}{rl}
& 1 \\
\hline
3a^2 = 3) & 60 \\
3a^2b = & 27 \\
\hline
& 334 \\
3ab^2 = & 243 \\
\hline
& 910 \\
b^3 = & 729 \\
\hline
& 1810 \\
& \text{etc.}
\end{array}
$$

According to §194, $\sqrt[3]{\tfrac{4}{5}}=\sqrt[3]{0.8}$ or $\sqrt[3]{\tfrac{4}{5}}=\sqrt[3]{\dfrac{4\cdot5\cdot5}{5\cdot5\cdot5}}=\dfrac{\sqrt[3]{100}}{5}$

The following examples should be worked out by the student:

$\sqrt[3]{731432701}=901$	$\sqrt[3]{1367631}=111$
$\sqrt[3]{351}=7.054004$	$\sqrt[3]{100}=4\cdot641588$
$\sqrt[3]{3}=1.44224$	$\sqrt[3]{\tfrac{3}{4}}=0.90855$
$\sqrt[3]{6\tfrac{3}{4}}=1.88207$	$\sqrt[3]{0.032768}=0.32$

In addition to the above exact methods of extracting roots, there is a simple method of approximation which may be made as accurate as desired, and is so valuable that it should be learned by everyone. It is based on the simple process of *factoring.*

The extraction of the square root of a number is simply the separating it into two *equal factors.* The extraction of the cube root of a number is the separation of it into three *equal factors,* etc., etc.

Now it is easy by inspection to tell quite closely into what two factors (for square root) a number may be divided, which shall be *nearly* equal. Thus the number 120 may be separated into the factors 2 and 60, but these are very unequal; or into the factors 8 and 15, which are much nearer, or into 10 and 12, which are quite near. Now taking the factors 10 and 12, we see that 10 is too small and 12 too large, and the desired factor must lie between them. So to find it approximately we add them together and divide by 2, thus $\frac{10+12}{2}=\frac{22}{2}=11$, and 11 is very nearly the true factor. This is the first approximation.

Now to get the root much closer we simply divide 120 by 11

to find out *exactly* what one factor must be if the other $=11$, and we have $\frac{120}{11}=10.90909\ldots$, so that the true root must lie somewhere between 10.90909 and 11. We therefore add these together and divide by 2, thus:

$$\frac{10.90909+11}{2}=10.9545\ldots$$

We find by actual calculation that the $\sqrt{120}=10.95445$, so that this simple method of inspection and approximation is as correct to within 0.00005 as the exact method. By taking the root above found and making a third approximation, the result would have been correct to many more places of decimals.

. The method may be put in the following rule :

1. Find by inspection of what two *nearly equal* factors the number is composed. The arithmetical mean of these two factors (i. e., one-half their sum) will be the *first* approximation.

2. Divide the number by this first approximation, carrying the division out to four or five decimal places. The arithmetical mean of the divisor and the quotient will be the *second* approximation.

This will usually give a correct answer to the fourth decimal place, but a third approximation, using the root last found as a divisor, will be found as accurate as any known process and can be relied on for any practical purpose whatever.

Another example will show the general character of the rule.

EXAMPLE. Find $\sqrt{346.285}$.

We see that $10^2=100$, is too small for the root, $20^2=400$ is much nearer, so we will try 20 for one factor and see what the other factor will be.

$$\frac{346.285}{20}=17.31425$$

Hence the number is composed of the two factors 20 and 17.31425. These are not greatly different from each other and their mean will give a fairly close approximation :

$$\frac{20+17.31425}{2}=18.657125$$

Now the error in this probably lies beyond the second decimal, so we take 18.65 as the divisor for the second approximation, thus:

$$\frac{346.285}{18.65}=18.5676$$

and the two factors are 18.65 and 18.5676, and their mean$=$

$$\frac{18.65+18.5676}{2}=18.6088$$

The true root is 18.60873, so that the closeness of the second approximation is seen.

Another example. Required $\sqrt{725892}$.

Now $1000^2=1000000$, which is too large. We try $800^2=640000$, and $900^2=810000$, and see that the root must lie between them, and that it is nearer 810000 than it is to 640000. We therefore take as our first trial divisor a number near 900, say 870.

$$\frac{725892}{870}=834$$

$$\frac{870+834}{2}=852=\text{first approximation.}$$

$$\frac{725892}{852}=851.9859$$

$$\frac{852+851.9859}{2}=851.99295=\text{second approxi-}$$

mation, correct to fifth decimal place.

A similar method is applicable to the extraction of the cube root, and it is one of the advantages of this system that it is as simple for the higher as for the lower roots.

To extract the cube root we have to separate the number into *three* equal factors. The rule may be at once stated, as follows:

1. Find by inspection of what three *nearly* equal factors the number is composed. The arithmetical mean of these (i. e., one-third their sum) will be the first approximation.

2. Divide the number by the first approximation, and the quotient again by the first approximation. The two divisors and the quotient are then three *very nearly* equal factors, and their arithmetical mean will be the required root, very closely. This is the *second* approximation.

A third approximation, using the second result as the divisor twice, and taking the mean of three new factors, will give a very accurate result, but this is rarely necessary.

EXAMPLES.

$\sqrt[3]{725892}=?$

$100^3=1000000$, so the root is something near 100. Let us try 90, for $90^3=729000$.

$$\frac{725892}{90}=8065$$

and $\frac{8065}{90}=89.6$

and the three factors are 90, 90 and 89.6, and their mean is:

$$\frac{90+90+89.6}{3}=89.86=\text{first approximation.}$$

Now $\frac{725892}{89.86}=8078.03$

and $\frac{8078.08}{89.86}=89.8957$

and we have for the three factors:

$$89.86, \quad 89.86, \quad \text{and} \quad 89.8957$$

and their mean will be:

$$\frac{89.86+89.86+89.8957}{3}=89.8719$$

for the second approximation, correct to the fourth decimal place.

A little practice will render the selection of the first trial root a very simple matter, and the first approximation may often be made mentally, so that the actual work is not laborious. The method is also very easily remembered and the principle is applicable to the 4th, 5th, or indeed any root. It will be good practice to apply this method to the foregoing examples and compare the results with those obtained by the exact, but more laborious, process.

BOOK XVI.

Calculations with Powers and Roots in General.

201.

When it is desired to extract the root of a power the following short method of indicating the operation is used. The exponent of the root is taken as the denominator and the exponent of the power as the numerator, to form a *fractional* exponent. Thus to indicate the nth root of a^m we may write $\sqrt[n]{a^m}$, and this would be perfectly correct, but it is better to write $a^{\frac{m}{n}}$. In this way we may write instead of $\sqrt[3]{8^2}$, more briefly $8^{\frac{2}{3}}$ (read 8 to the two-thirds power). Likewise we have $\sqrt{a^3}=a^{\frac{3}{2}}$. Every quantity which has no exponent is considered to be its first power and may be so written; thus $a=a^1$; whence $\sqrt[3]{a}=\sqrt[3]{a^1}=a^{\frac{1}{3}}$; $\sqrt{a}=a^{\frac{1}{2}}$.

Conversely we have: $x^{\frac{m}{n}}=\sqrt[n]{x^m}$; $x^{\frac{1}{n}}=\sqrt[n]{x}$; $5^{\frac{1}{3}}=\sqrt[3]{5}$, etc. This use of fractional exponents, due to Descartes, greatly simplifies the subject of calculations with roots and powers, and enables the operations to be much more clearly understood.

202.

When a quantity is written with a fractional exponent, it means that the quantity is divided into as many equal factors as there are units in the denominator, and that one of these factors is taken as many times as there are units in the denominator.

It is evidently a matter of no consequence, in calculating with fractional exponents, whether the root is first extracted and the quantity then raised to the required power, or whether the power is first raised and then the root extracted. For instance:

$$8^{\frac{2}{3}}=\sqrt[3]{8^2}=\sqrt[3]{64}=4$$

which is just the same as :

$$8^{\frac{2}{3}}=(\sqrt[3]{8})^2=2^2=4$$

and in general :

$$a^{\frac{m}{n}}=\sqrt[n]{(a^m)}=(\sqrt[n]{a})^m$$

The correctness of this statement, so important in calculating with powers, may be demonstrated in the following manner :

If we raise a given power, such as a^3, again to another power, the 4th for example, we indicate this operation thus $(a^3)^4$, and we shall obtain a power of a higher degree, equal to that of the product of the two exponents. This is evident from the fact that we are taking the product of three equal factors, $aaa=a^3$, four times again, as a factor, thus $aaa·aaa·aaa·aaa$, which gives $3·4=12$ equal factors. or $(a^3)^4=a^{12}$. But we see that $(a^3)^4$ will thus be the same as $(a^4)^3$, or in general $(a^n)^m=(a^m)^n$.

Now in order to show that in general

$$(\sqrt[n]{a})^m=\sqrt[n]{(a^m)}$$

we may consider the quantity a to be separated into n equal factors, or $a=w^n$. Substituting w^n for a, we have :

$$(\sqrt[n]{a})^m=(\sqrt[n]{w^n})^m=w^m$$

and also

$$\sqrt[n]{(a^m)}=\sqrt[n]{(w^n)^m}=\sqrt[n]{(w^m)^n}=w^m$$

whence the general statement follows :

$$(\sqrt[n]{a})^m=\sqrt[n]{(a^m)}$$

203.

When the numerator and denominator of a fractional exponent, or, what is the same thing, the exponents of the power and the root, are both multiplied or divided by the same quantity the value of the expression remains unchanged.

Thus we have :

$$64^{\frac{2}{3}}=64^{\frac{4}{6}}$$
$$\sqrt[3]{64^2}=\sqrt[6]{64^4}$$

In general :

$$a^{\frac{m}{n}} = a^{\frac{mp}{np}}$$

$$\sqrt[n]{a^m} = \sqrt[np]{a^{mp}}$$

If, therefore, the quantity a is divided into p equal factors, and one of these factors again taken p times, we must obviously get the same result again.   Thus, for example :

$$(\sqrt[3]{64})^2 = (\sqrt[3]{4 \cdot 4 \cdot 4})^2 = 4^2 = (2 \cdot 2)^2 = 2^4$$

and, $\quad (\sqrt[6]{64})^4 = (\sqrt[6]{2 \cdot 2 \cdot 2 \cdot 2 \cdot 2 \cdot 2})^4 = 2^4$

In general, since according to § 202, we may consider the quantity a as separated into np factors and w^{np} placed instead of a, we have :

$$(\sqrt[n]{a})^m = (\sqrt[n]{w^{np}})^m = (w^p)^m = w^{mp}$$

and, $\quad (\sqrt[np]{a})^{mp} = (\sqrt[np]{w^{np}})^{mp} = w^{mp}$

By the application of these principles a number of fractional exponents may be brought to a common denominator, and as will be seen in the following pages, many expressions very much simplified.

These principles may also be used to combine a number of roots together without knowing their actual values, and also to distinguish which of several values, such as $\sqrt{3}$, $\sqrt[3]{5}$, $\sqrt[6]{24}$, is the greatest or least.   If we write these with fractional exponents, instead of using the root symbols, we have :

$$\sqrt{3} = 3^{\frac{1}{2}} = 3^{\frac{3}{6}} = \sqrt[6]{3^3} = \sqrt[6]{27}$$

$$\sqrt[3]{5} = 5^{\frac{1}{3}} = 5^{\frac{2}{6}} = \sqrt[6]{5^2} = \sqrt[6]{25}$$

$$\sqrt[6]{24} = 24^{\frac{1}{6}} = 24^{\frac{1}{6}} = \sqrt[6]{24}$$

and we see at once that $\sqrt{3}$ is the greatest, and $\sqrt[6]{24}$ is the least of the three quantities.

204.

In order to multiply any powers of one base (or root) together it is only necessary to *add* the exponents.   That is, the product of two powers of the same base is a power of that base of which the exponent is equal to the sum of the exponents of the two factors. This is evident, when we consider that the product must consist of as many equal factors as there are in both multiplier and multiplicand taken together, thus :

$$a^5.a^2 = a^{5+2} = a^7 ;$$

We see the truth of this, because :

$$a^5.a^2 = aaaaa.aa = a^7$$

Likewise :

$$3^4 + 3^2 + 3^5 = 3^{22} ;$$
$$x^5 x^3 x = x^{5+3+1} = x^9$$
$$b^{15} b^{15} = b^{30}$$

Also when the exponents are fractional we have the same rule :

$$a^{\frac{3}{7}} \cdot a^{\frac{2}{7}} = a^{\frac{3}{7}+\frac{2}{7}} = a^{\frac{5}{7}} ;$$

since the one contains the 7th root of a, twice, and their product contains it five times as a factor.

EXAMPLES :

$$a^m a^n = a^{m+n} ;$$

$$a^m a^n a^p = a^{m+n+p} ;$$

$$x^{\frac{m}{n}} x^{\frac{p}{n}} = x^{\frac{m}{n}+\frac{p}{n}} = x^{\frac{m+p}{n}}$$

$$x^{\frac{n}{3}} x^{\frac{n}{3}} = x^{\frac{2n}{3}} = x^n$$

$$x^m x = x^{m+1}$$

$$x^{\frac{m}{n}} x = x^{\frac{m}{n}+1} = x^{\frac{m+n}{n}}$$

$$2a^5 b^3 \cdot 3^2 a^2 b = 2 \cdot 3 \cdot a^5 \cdot a^2 \cdot b^3 \cdot b = 6a^7 b^4 \quad (\S\,89,\ 1)$$

$$a^4 (a^3 - a^2 + 1) = a^7 - a^6 + a^4 \quad (\S\,89,\ 2)$$

$$3x^3 (2x^4 + 4x + 3) = 6x^9 + 12x^6 + 9x^5$$

$$a^2 b^4 (a^2 b + ab^3 + b^4) = a^4 b^5 + a^3 b^7 + a^2 b^8$$

$$(a-1)(a^3 + a^2 + a + 1) = a^4 - 1 \quad (\S\,89,\ 3)$$

$$(a-b)(a^3 + a^2 b + ab^2 + b^3) = a^4 - b^4$$

$$(a-b)(a^2 + 2ab + b^2) = a^3 - b^3$$

$$(a^2 + b^2)(a^2 - b^2) = a^4 - b^4 \quad (\S\,91)$$

The working out of the above examples will furnish the student with good practice in reduction, and none of them should be passed over if not understood.

205.

In order to divide one power in a given base by another power of the same base, it is only necessary to *subtract* the exponent of the divisor from the exponent of the dividend.

This is the converse of the rule for multiplication, and the reason will readily be seen when we consider that any number of common factors in the divisor cancel the same number in the dividend. Thus we have:

$$\frac{8^7}{8^4}=8^{7-4}=8^3, \text{ for } \frac{8^7}{8^4}=\frac{8\cdot8\cdot8\cdot8\cdot8\cdot8\cdot8}{8\cdot8\cdot8\cdot8}=8\cdot8\cdot8=8^3$$

also:

$$\frac{a^{\frac{5}{7}}}{a^{\frac{3}{7}}}=a^{\frac{5}{7}-\frac{3}{7}}=a^{\frac{2}{7}} \quad \text{(see § 203)}$$

because:

$$\frac{a^{\frac{5}{7}}}{a^{\frac{3}{7}}}=\frac{(\sqrt[7]{a})^5}{(\sqrt[7]{a})^3}=(\sqrt[7]{a})^{5-3}=(\sqrt[7]{a})^2=a^{\frac{2}{7}}$$

likewise:

$$\frac{x^{\frac{3}{4}}}{x^{\frac{2}{3}}}=x^{\frac{1}{12}}$$

When the exponent of the divisor is greater than that of the dividend, the difference of their exponents is taken and the exponent of the quotient is negative. i. e., has the minus sign prefixed;

Thus we have:

$$\frac{a^4}{a^7}=a^{4-7}=a^{-3}; \quad \frac{a^{\frac{2}{3}}}{a^{\frac{3}{4}}}=a^{-\frac{1}{12}}$$

When therefore a negative exponent results from the division of one power of a root by another power of the same root, it simply indicates that the divisor consists of as many more factors than the dividend, as there are units in the negative exponent.

A quantity with a negative exponent is therefore *not* itself negative, but is equal to unity divided by the same quantity with a positive exponent, thus:

$$a^{-3}=\frac{1}{a^3}; \qquad x^{-\frac{1}{12}}=\frac{1}{x^{\frac{1}{12}}}; \qquad x^{-n}=\frac{1}{x^n};$$

$$1^{-1}=\frac{1}{1}=1; \qquad\qquad 16^{-\frac{1}{2}}=\frac{1}{16^{\frac{1}{2}}}=\frac{1}{\sqrt{16}}=\frac{1}{4}.$$

This enables us to transfer a quantity from the denominator to the other numerator of a fraction, when so desired, merely by changing the sign of its exponent, thus, instead of :

$$\frac{a^3 x^4}{y^5} \text{ we may write } a^3 x^4 y^{-5}$$

If the dividend and divisor are equal to each other, the quotient is always equal to unity. Thus :

$$\frac{a^7}{a^7} = 1 \; ; \; \frac{x^n}{x^n} = 1$$

But the above rule for division gives in such cases the exponent o.

$$\frac{a^7}{a^7} = a^{7-7} = a^0 \; ; \; \frac{x^n}{x^n} = x^{n-n} = 1$$

A quantity with zero for an exponent is therefore always equal to 1, and the mistake must not be made of supposing it to to be equal to o. Thus, we have :

$$a^0 = 1 \; ; \; x^0 = 1 \; ; \; \left(\frac{a}{b}\right)^0 = 1 \; ; \; 1^0 = 1$$

206.

Since the rules given in the preceding paragraphs are quite general in their applications, they must apply as well to negative as to positive exponents, and this we see to be true from the following examples :

MULTIPLICATION.

$$a^7 \cdot a^{-4} = a^{7-4} = a^3, \text{ because :}$$

$$a^7 \cdot a^{-4} = a^7 \cdot \frac{1}{a^4} = a^3$$

$$a^{-3} \cdot a^{-2} = a^{-3-2} = a^{-5}, \text{ because :}$$

$$a^{-3} \cdot a^{-2} = \frac{1}{a^3} \cdot \frac{1}{a^2} = \frac{1}{a^5} = a^{-5}$$

DIVISION.

$$\frac{a^7}{a^{-4}} = a^{7-(-4)} = a^{11}, \text{ because :}$$

$$\frac{a^7}{a^{-4}} = a^7 : \frac{1}{a^4} = a^7 \cdot a^4 = a^{11}$$

$$\frac{a^{-4}}{a^{-7}} = a^{-4-(-7)} = a^3, \text{ because :}$$

$$\frac{a^{-4}}{a^{-7}} = \frac{1}{a^4} : \frac{1}{a^7} = \frac{1}{a^4} \cdot \frac{a^7}{1} = a^3$$

GENERAL EXAMPLES:

$a^m a^{-n} = a^{m-n}$

$\dfrac{a^{-n}}{a^{-m}} = a^{-n+m} = a^{m-n}$

$a^{m-n} \cdot a^n = a^m$

$\dfrac{a^m}{a^{2m}} = a^{-m}$

$\dfrac{a^n}{a} = a^{n-1}$

$\dfrac{a^m}{a^{m-n}} = a^{m-m+n} = a^n$

$\dfrac{4a^5 b^7 c^3}{2ab^4 c^3} = 2a^4 b^3$

$a^m a^{-1} = a^{m-1}$

$\dfrac{a^{\frac{m}{n}}}{a} = a^{\frac{m}{n}-1} = a^{\frac{m-n}{n}}$

$a^{m-1} \cdot a^{1-m} = a^0 = 1$

$\dfrac{a}{a^n} = a^{1-n}$

$\dfrac{a}{a^{\frac{m}{n}}} = a^{1-\frac{m}{n}} = a^{\frac{n-m}{n}}$

$2x^6(3x - 4x^3) = 6x^7 - 8x^9$

$\dfrac{6a^5}{9a^9} = \dfrac{2}{3a^4} = \tfrac{2}{3}a^{-4}$

$$(2a^3 b^5 - 3ab^{-4})(5a^{-2}b^3 + ab^4) = 10ab^8 + 2a^4 b^9 - 15a^{-1}b^{-1} - 3a^2$$

207.

In order to raise any power to a higher power, multiply the xponent by the new exponent. Thus, to raise $(a^4)^3$, we have:

$$(a^4)^3 = a^{12}, \text{ because :}$$
$$(a^4)^3 = a^4 \; a^4 \cdot a^4 = a^{4+4+4} = a^{4 \cdot 3} = a^{12}$$

Likewise:

$$(a^{-3})^2 = a^{-6}, \text{ for}$$
$$(a^{-3})^2 = a^{-3} \cdot = \frac{1}{a^3} \cdot \frac{1}{a^3} = \frac{1}{a^6} = a^{-6}$$

$$(a^{\frac{3}{8}})^2 = a^{\frac{6}{8}} = a^{\frac{3}{4}}, \text{ because :}$$
$$(a^{\frac{3}{8}})^2 = [(\sqrt[8]{a})^3]^2 = (\sqrt[8]{a})^6 = a^{\frac{6}{8}}$$

In general therefore, we have:

$$(a^m)^n = a^{mn}$$
$$[(a^m)^n]^p = a^{mnp}$$
$$\left(a^{\frac{m}{n}}\right)^n = a^m$$
$$\left(\frac{a^m}{b^n}\right)^n = \frac{a^{mn}}{b^{nn}}$$
$$\left(\frac{a}{b}\right)^{-n} = \frac{b^n}{a^n}$$
$$\left(a^{\frac{2}{8}}\right)^3 = a^2$$

208.

If any root is to be extracted from any given power, the exponent of the power is to be divided by the exponent of the required root. If the division cannot be performed it is to be indicated.

Thus :

$$\sqrt[3]{2^9}=2^{\frac{9}{3}}=2^3 ; \qquad \sqrt[4]{4^{\frac{3}{2}}}=4^{\frac{1}{2}} ;$$

$$\sqrt{5^{\frac{3}{8}}}=5^{1\frac{3}{6}} ; \qquad \sqrt{a^3}=a^{\frac{3}{2}} ;$$

$$\sqrt[4]{a^{\frac{5}{6}}}=a^{\frac{2}{7}} ; \qquad \sqrt[5]{a^{15}}=a^3 .$$

In general :

$$\sqrt[n]{a^m}=a^{\frac{m}{n}} ; \qquad \sqrt[n]{a^{m+1}}=a^{\frac{m+1}{n}} ;$$

$$\sqrt[n]{a^{-m}}=a^{-\frac{m}{n}} ; \qquad \sqrt[n]{a^{m-n}}=a^{\frac{m-n}{n}} ;$$

$$\sqrt[n]{a^{mn}}=a^m ; \qquad \sqrt{a^{2n}}=a^{\frac{2n}{2}}=a^n ;$$

$$\sqrt[n]{a^{2n}}=a^2 ; \qquad \sqrt{a^{-4}}=a^{-2} .$$

209.

When it is required to raise the product of any factors to given power, each factor is raised to the required power separarely. Thus, for example :

$(2\cdot3)^3=2^3\cdot3^3$, for we have :

$(2\cdot3)^3=2\cdot3\cdot2\cdot3\cdot2\cdot3=2\cdot2\cdot2\cdot3\cdot3\cdot3=2^3\cdot3^3.$

In general $(abc)^n=a^nb^nc^n$

EXAMPLES :

$$(a^2b^5)^3=a^6b^{15} ; \qquad (\tfrac{1}{2}a)^2=\tfrac{1}{4}a^2;$$

$$(\tfrac{1}{3}\sqrt{a})^2=\tfrac{1}{9}a ; \qquad (\tfrac{2}{3}\sqrt{a})^2=\tfrac{4}{9}a;$$

$$(\tfrac{1}{5}\sqrt[3]{x})^3=\tfrac{x}{125} ; \qquad (\sqrt[n]{a}\cdot\sqrt[n]{b})^n=ab$$

210.

Conversely, if it is required to extract a root of a quantity composed of several factors, the root of each factor is extracted separately.

When the radical sign $\sqrt{\ }$ stands before a parenthesis containing factors, the parenthesis may be dropped as unnecessary, the sign being understood to apply to all the factors.

Instead of $\sqrt{(abc)}$, we may write more briefly $\sqrt{abc}$.

$$\sqrt[n]{a^2b^6c^9}=\sqrt[n]{a^2}\cdot\sqrt[n]{b^6}\cdot\sqrt[n]{c^9}=ab^3c^3$$

In general $\sqrt[n]{ab}=\sqrt[n]{a}\cdot\sqrt[n]{b}$

The truth of these statements is apparent from what has preceded.

A quantity, which, when raised to the nth power, gives the quantity ab, is the nth root of ab. Now since, according to § 209, $(\sqrt[n]{a}\cdot\sqrt[n]{b})=ab$, so, conversely $\sqrt[n]{ab}=\sqrt[n]{a}\cdot\sqrt[n]{b}$.

EXAMPLES :

$$\sqrt{\frac{a^2b^4}{c^6x^2}}=\frac{ab^2}{c^3x}$$

$$\sqrt[3]{\frac{27a^3}{8b^6}}=\frac{3a}{2b^2}$$

211.

(1) If we can separate any quantity under a radical sign, into two such factors that one of them shall have a rational root, we may extract the root of this factor and place it outside the radical sign as a coefficient of the other factor still under the radical sign. Thus, for instance :

$$\sqrt{45}=\sqrt{9\cdot5}=3\sqrt{5}$$

In this way expressions containing roots, may often be simplified and collected.

The term *radical* is often applied to a quantity of which a root is to be extracted, the term applying to the whole expression including the radical sign and the quantity under it. Thus :

$\sqrt[3]{a}$; $\sqrt{a}$; $a^{\frac{3}{4}}$; $\sqrt[4]{a^3}$; $\sqrt{45}$; $\sqrt{27}$, etc., may all be called radicals.

Radicals are said to be *similar* when they have the same exponents and the same quantities under the radical sign, no matter how different the coefficients may be. Thus :

$\sqrt{2}$; $3\sqrt{2}$; $\frac{1}{2}\sqrt{2}$ are similar, as also are $\sqrt{b}$, and $3\sqrt{b}$, or $\sqrt{ab}$ and $2\sqrt{ab}$: but $2\sqrt{2}$; $2\sqrt[3]{2}$; or $\sqrt{ab}$ and $\sqrt[3]{ab}$, are dissimilar.

EXAMPLES :

$\sqrt{18}=\sqrt{9\cdot2}=\sqrt{9}\sqrt{2}=3\sqrt{2}$;

$\sqrt{8}=\sqrt{4\cdot2}=\sqrt{4}\sqrt{2}=2\sqrt{2}$;

$\sqrt{a^2b}=\sqrt{a}\sqrt{b}$;

$\sqrt[3]{a^3b}=a\sqrt[3]{b}$;

$$\sqrt{a^4b^5c^2}=\sqrt{a^4b^4c^2b}=a^2b^2c\sqrt{b} \; ;$$
$$\sqrt[3]{24a^7b}=\sqrt[3]{8\cdot3\cdot a^6ab}=2a^2\sqrt[3]{3ab} \; ;$$
$$2\sqrt{18x^3y^5}=2\sqrt{9\cdot2x^2\cdot y^4\cdot y}=6xy^2\sqrt{2y} \; ;$$
$$\sqrt[n]{a^{m+n}b^{2n}}=\sqrt[n]{a^ma^nb^{2n}}=ab^2\sqrt[n]{a^m}\cdot$$

(2) Conversely, factors without the radical sign, may be placed under it, if they are first raised to the power indicated by its exponent.

Thus, for example :

$$3\sqrt{2}=\sqrt{3^2\cdot2}=\sqrt{18}$$
$$2\sqrt{2}=\sqrt{4\cdot2}=\sqrt{8}$$
$$\tfrac{1}{2}\sqrt{2}=\sqrt{\tfrac{1}{4}\cdot2}=\sqrt{\tfrac{1}{2}}$$
$$a\sqrt{b}=\sqrt{a^2b}$$
$$a\sqrt[3]{b}=\sqrt[3]{a^3b}$$
$$x\sqrt[n]{x}=\sqrt[n]{x^{n+1}}$$

Calculations With Roots.

212.

It is to be noted that the simplicity of an expression does not depend upon the *degree* of the exponents, but upon the fewness of its members and radical signs. When calculations are actually made with powers and roots, the operations are performed by means of logarithms, (as will shortly be explained) and in such cases the extraction of the 100th root involves no more labor than the extraction of the square root.

(1) *Addition and Subtraction of Roots.* If the radicals are *dissimilar*, the operations can only be indicated, but if they are similar, or can be made similar we have only to add or subtract the coefficients, as the case may be. Thus :

$$\sqrt{5}+\sqrt{6}=\sqrt{5}+\sqrt{6} \; ; \qquad 2\sqrt[3]{a}-3\sqrt[3]{a}=2\sqrt[3]{a}-3\sqrt[3]{a} \; ;$$
$$2\sqrt{2}+3\sqrt{2}=5\sqrt{2} \; ; \qquad 2a^{\frac{2}{3}}-5a^{\frac{2}{3}}+6a^{\frac{2}{3}}=3a^{\frac{2}{3}} \; ;$$
$$a\sqrt{h}-b\sqrt{h}=(a-b)+\sqrt{h} \; ; \quad ax^{\frac{m}{n}}-bx^{\frac{m}{n}}=(a-b)x^{\frac{m}{n}} \; ;$$
$$\sqrt{8}+\sqrt{18}=2\sqrt{2}+3\sqrt{2}=5\sqrt{2}=\sqrt{50} \; ; \text{ see § 211,2} \; ;$$
$$2\sqrt{27}-3\sqrt{12}=2\sqrt{9\cdot3}-3\sqrt{4\cdot3}=6\sqrt{3}-6\sqrt{3}=0 \; ;$$
$$\sqrt[3]{24}+\sqrt[3]{3}=2\sqrt[3]{3}+\sqrt[3]{3}=3\sqrt[3]{3}=\sqrt[3]{81}.$$

(2) *Multiplication of Roots.* The roots are first to be reduced to the same degree (i. e., same exponent, see § 203). They

can then be brought under the same radical sign, and the similar factors collected. The other factors without the radical must be multiplied together separately and their product placed outside the radical sign.

EXAMPLES :

$\sqrt{a} \cdot \sqrt{b} = \sqrt{ab}$;　　　　　　$\sqrt[3]{a} \cdot \sqrt[3]{b} = \sqrt[3]{ab}$;

$\sqrt{3} \cdot \sqrt{12} = 6$;　　　　　　$2\sqrt{a} \cdot 3\sqrt{b} = 6\sqrt{ab}$;

$\sqrt{a} \cdot \sqrt{a} = a$;　　　　　　$a\sqrt{b} \cdot b\sqrt{a} = ab\sqrt{ab}$;

$\sqrt{a} \cdot \sqrt[3]{b} = a^{\frac{1}{2}} \cdot b^{\frac{1}{3}} = a^{\frac{3}{6}} \cdot b^{\frac{2}{6}} = \sqrt[6]{a^3 b^2}$;

$2\sqrt{a^2 b} \cdot 3\sqrt{ab} = 6a^2 b$;

$3\sqrt{a} \cdot 5\sqrt[3]{b} = 15\sqrt[6]{a^3 b^2}$;

$\sqrt[n]{x} \cdot \sqrt[n]{x^m} = \sqrt[n]{x^{m+1}}$;

$3\sqrt{5} \cdot \sqrt{5} = 15$;

$a\sqrt{(\sqrt{a} + \sqrt{b})} = a + \sqrt{ab}$;

$(\sqrt{a} + \sqrt{b})(\sqrt{a} - \sqrt{b}) = ab$;　(see § 180,2).

$(\sqrt{x} + \sqrt{y})(\sqrt{x} + \sqrt{y}) = (\sqrt{x} + \sqrt{y})^2 = x + 2\sqrt{xy} + y$;

(See § 186).

(3) *Division.* If the exponents of the radicals are alike, or can be made alike, the quantities can be immediately placed under the same radical sign. Thus :

$\dfrac{\sqrt{a}}{\sqrt{b}} = \sqrt{\dfrac{a}{b}}$;　　　　　　$\dfrac{\sqrt[3]{a}}{\sqrt{b}} = \sqrt[6]{\dfrac{a^2}{b^3}}$;

$\dfrac{\sqrt{a}}{\sqrt{a}} = 1$;　(see § 180, 2)　　$\dfrac{\sqrt{8}}{\sqrt{18}} = \sqrt{\dfrac{8}{18}} = \sqrt{\dfrac{4}{9}} = \dfrac{2}{3}$;

$\dfrac{2\sqrt{3}}{\sqrt{12}} = \dfrac{\sqrt{4 \cdot 3}}{\sqrt{12}} = 1$;　　$\dfrac{2\sqrt{27}}{3\sqrt{12}} = \dfrac{2}{3}\sqrt{\dfrac{27}{12}} = \dfrac{2}{3} \cdot \sqrt{\dfrac{9}{4}} = \dfrac{2}{3} \cdot \dfrac{3}{2} = 1$;

$\dfrac{\sqrt{45} - \sqrt{20}}{\sqrt{5}} = \sqrt{\dfrac{45}{5}} - \sqrt{\dfrac{20}{5}} = 3 - 2 = 1$.

213.

The general problem, of the raising of any polynomial quantity to any power, and conversely, the extraction of any root, belongs properly to the subject of Analysis, or Higher Algebra, where it can readily be solved by means of Newton's Binomial Theorem.

For the present it will be sufficient to give the rules for the formation of the second and third powers. For the formation of the second power, the following simple law is applied.

Suppose $a+b+c+d$.., be any polynomial quantity. If we square this by actual polyplication, and write the successive products under each other, we have:

$$
\left.
\begin{array}{l}
a+b+c+d+e+\ldots \\
a+b+c+d+e+\ldots
\end{array}
\right\} \text{Factors}
$$

$$
\begin{array}{l}
a^2+ab+ac+ad+ae+\ldots \\
ab+b^2+bc+bd+be+\ldots \\
ac+bc+c^2+cd+ce+\ldots \\
ad+bd+cd+d^2+de+\ldots \\
ae+be+ce+de+e^2+\ldots
\end{array}
$$

Comparing the successive lines we perceive by inspection the simple rule by which the square of any polynomial is obtained, namely: *The square of any polynomial quantity consists of the squares of each member and the double product of each member by each of the others.* If any members have minus signs it must be remembered that like signs produce plus, and unlike signs minus, and that the squares are therefore always positive. Thus:

$$(a+b+c)^2=a^2+b^2+c^2+2ab+2ac+2bc \ ;$$
$$(a-b-c)^2=a^2+b^2+c^2-2ab-2ac+2bc \ ;$$
$$(a+b)^2=a^2+2ab+b^2 \ ; \qquad (a-b)^2=a^2-2ab+b^2 \ ;$$
$$(x+a)^2=x^2+2ax+a^2 \ ; \qquad (x-1)^2=x^2-2x+1 \ ;$$
$$(1-x)^2=1-2x+x^2 \ ; \qquad (x+\tfrac{1}{2})^2=x^2+x+\tfrac{1}{4} \ ;$$
$$(x+\tfrac{p}{2})^2=x^2+px+\tfrac{p^2}{4} \ ; \qquad (x-\tfrac{p}{2})^2=x^2-px+\tfrac{p^2}{4} \ ;$$
$$(x-\tfrac{2}{3}a)^2=x^2-\tfrac{4}{3}ax+\tfrac{4}{9}a^2 \ ; \quad (3ax+b)^2=9a^2x^2+6abx+b^2 \ ;$$
$$(x^m+a)^2=x^{2m}+2ax^m+a^2 \ ; \quad (\sqrt{}(x-\sqrt{}y)^2x-2\sqrt{}xy+y \ ;$$
$$x^2-(a-x)^2=x^2-(a^2+2ax+x^2)=2ax-a^2 \ ;$$
$$b^2-(a-c)^2=[b+(a-c)] \ [b-(a-c)]=(b+a-c) \ (b-a+c).$$

(See § 91.)

(2) In the same way we may derive a general law for the formation of the cube. The principle is the same, the square multiplied again by the base giving the cube, but the work is much more prolix, and as the only cube used in elementary

algebra is that of a binomial, we will here simply repeat the rule given in § 198, calling the attention of the student to the arrangement of the plus and minus signs, which had better be memorized.

$$(x+y)^3 = x^3 + 3x^2y + 3xy^2 + y^3 \ ;$$
$$(a-b)^3 = a^3 - 3a^2b + 3ab^2 - b^3 \ ;$$
$$(a^m - b^n)^3 = a^{3m} - 3a^{2m}b^n + 3a^m b^{2n} - b^{3n}.$$

214.

The rules for the converse problems, i. e., the extraction of the square and cube roots of polynomial quantities, follow from the preceding. An experienced student will rarely have occasion to use a rule, but will detect a root which can be extracted at the first glance. It is not often that the root can be exactly extracted, and the operation must generally be indicated, either by enclosing the expression in a parenthesis, preceded by the radical sign or followed by a fractional exponent, and treating the quantity as if it were a monomial.

For instance, the square root of the binomial $a+b$, is simply *indicated*, thus $\sqrt{(a+b)}$, or $\sqrt{a+b}$, or $(a+b)^{\frac{1}{2}}$. Beginners must be careful not to confuse $\sqrt{(a+b)}$ with $\sqrt{a}+\sqrt{b}$, or $\sqrt{(a-b)}$ with $\sqrt{a}-\sqrt{b}$, these expressions being by no means the same. as will be seen by the following numerical examples.

$$\sqrt{16}+\sqrt{9}=7, \text{ while } \sqrt{(16+9}=5.)$$

Also :

$$16+\sqrt{9}=19, \text{ while } \sqrt{16+9}=13.$$

Again :

$$\sqrt{25}-\sqrt{16}=1, \text{ while } \sqrt{(25-16)}=3.$$

The following points should be noted in regard to the extraction of roots ; every power of a *monomial* quantity is a monomial, but the square of a binomial quantity consists of three parts of which two are perfect and positive squares, while the cube of a binomial quantity consists of four parts of which two are perfect cubes. It therefore follows that no square root of a binominal quantity is possible, and that the square root of a trinomial quantity when a root is possible, must always be a binomial. To find this root, take the square roots of the two perfect squares of the trinomial and join them by the sign of the

third member. If this expression squared gives the original trinomial expression, it is the desired root; if it does not, the quantity has no rational square root.

A similar rule can be formulated for cube root. (Compare §
324).

EXAMPLES :

$$\sqrt{x^2+2xy+y^2}=\sqrt{(x+y)^2}=x+y$$

$$\sqrt{9x^2-6xy+y^2}=\sqrt{(3x-y)^2}=3x-y$$

$$\sqrt{x^2+px+\tfrac{p^2}{4}}=\sqrt{(x+\tfrac{p}{2})^2}=x+\tfrac{p}{2}$$

$$\sqrt{(1-2x+x^2)}=\sqrt{(1-x)^2}=1-x$$

$$\sqrt{x^2-\tfrac{4}{3}x+\tfrac{4}{9}}=\sqrt{(x-\tfrac{2}{3})^2}=x-\tfrac{2}{3}$$

Incomplete
Squares

$$\sqrt{a^2+4ax+x^2}=\sqrt{a^2+4ax+x^2}$$

$$\sqrt{a^2+b^2}=\sqrt{a^2+b^2}$$

$$\sqrt{a^2-b^2}=\sqrt{(a+b)(a-b)}$$

$$\sqrt{(a+b)^2(a-b)^2}=(a+b)(a-b)=a^2-b^2$$

$$\sqrt{(x+y)^3}=\sqrt{(x+y)^2(x+y)}=(x+y)\sqrt{x+y}$$

$$\sqrt[3]{(x^3+3x^2y+3xy^2+y^3)}=\sqrt[3]{(x+y)^3}=x+y$$

$$\sqrt[3]{(a^3-3a^2x+3ax^2-x^3)}=\sqrt[3]{(a-x)^3}=a-x$$

$$\sqrt[3]{(a^3+b^3)}=(a^3+b^3)^{\tfrac{1}{3}}$$

$$\sqrt[3]{(a^3-b^3)}=(a^3-b^3)^{\tfrac{1}{3}}$$

Incomplete
Cubes

For the extraction of the square root of polynomial
expressions see § 324.

215.

The following reductions and changes of form will often be
found useful, and should be carefully studied :

(1) When the denominator of a fraction consists of a single
member under a radical sign, this quantity may be made rational
by multiplying both numerator and denominator by such fractional power of the denominator as is required to make the
power complete. Thus :

$$\frac{\sqrt[3]{3}}{\sqrt{2}}=\frac{\sqrt[3]{3}\cdot\sqrt{2}}{\sqrt{2}\cdot\sqrt{2}}=\frac{\sqrt[6]{3^2}\cdot\sqrt[6]{2^3}}{2}=\tfrac{1}{2}\sqrt[6]{72};$$

$$\frac{3}{\sqrt{3}} = \frac{3 \cdot \sqrt{3}}{\sqrt{3} \cdot \sqrt{3}} = \frac{3\sqrt{3}}{3} = \sqrt{3} \; ;$$

$$\frac{a}{\sqrt[3]{b}} = \frac{a \cdot b^{\frac{2}{3}}}{b^{\frac{1}{3}} b^{\frac{2}{3}}} = \frac{a}{b} \sqrt[3]{b^2} \; ;$$

$$\frac{2}{\sqrt{2}} = \frac{2\sqrt{2}}{\sqrt{2} \cdot \sqrt{2}} = \sqrt{2} \; ;$$

$$\frac{a}{\sqrt{a}} = \sqrt{a} \; ;$$

$$\frac{a}{\sqrt[3]{a}} = \frac{a^1}{a^{\frac{1}{3}}} = a^{\frac{2}{3}} = \sqrt[3]{a^2} \; ;$$

$$\frac{\sqrt{a+x}}{\sqrt{a-x}} = \frac{\sqrt{a+x} \cdot \sqrt{a-x}}{\sqrt{a-x} \cdot \sqrt{a-x}} = \frac{\sqrt{(a+x)(a-x)}}{a-x} = \frac{\sqrt{a^2+x^2}}{a-x} .$$

(2). When the denominator of the fraction is composed of two radicals the sign of one member must be changed. (§ 91).

EXAMPLES:

$$\frac{\sqrt{a}+\sqrt{x}}{\sqrt{a}-\sqrt{b}} = \frac{(\sqrt{a}+\sqrt{x})(\sqrt{a}+\sqrt{b})}{(\sqrt{a}-\sqrt{b})(\sqrt{a}+\sqrt{b})} = \frac{(\sqrt{a}+\sqrt{x})(\sqrt{a}+\sqrt{b})}{a-b} \; ;$$

$$\frac{x}{\sqrt{x}+\sqrt{y}} = \frac{x(\sqrt{x}-\sqrt{y})}{(\sqrt{x}+x\sqrt{y})(\sqrt{x}-\sqrt{y})} = \frac{x(\sqrt{x}-\sqrt{y}}{x-y}$$

$$\frac{x}{a+\sqrt{x}} = \frac{x(a-\sqrt{x})}{(a+\sqrt{x})(a-\sqrt{x})} = \frac{x(a-\sqrt{x})}{a^2-x}$$

(3). Radical quantities can always be reduced to a common denominator, thus:

$$\sqrt{ax} - \frac{ax}{a+\sqrt{x}} = \frac{(a+\sqrt{ax})\sqrt{ax}-ax}{a+\sqrt{x}} = \frac{a\sqrt{ax}}{a+\sqrt{ax}} \; ;$$

$$\frac{1}{(1-x^2)^{\frac{1}{2}}} + \frac{x^2}{(1-x^2)^{\frac{3}{2}}} = \frac{1}{(1-x^2)^{\frac{3}{2}}} = (1-x^2)^{-\frac{3}{2}}$$

(4). Miscellaneous Reductions:

$$\frac{\sqrt{9y^2(a^2-x^2)}}{y\sqrt{a+x}} = \frac{3y\sqrt{(a+x)(a-x)}}{y\sqrt{a+x}} = 3\sqrt{a-x} \; ;$$

$$\frac{x+y}{x-y}\sqrt{\frac{x-y}{x+y}} = \sqrt{\frac{(x-y)(x+y)^2}{(x-y)^2(x+y)}} = \sqrt{\frac{x+y}{x-y}} \; ;$$

$$(a+x)^m(a+x)^n = (a+x)^{m-n} \; ;$$

$$\frac{(a+x)^m}{(a+x)^n}=(a+x)^{m-n}\ ;$$

$$a^m\left(\mathbf{1}+\frac{x^m}{a^m}\right)=a^m+x^m\ ;$$

$$a^m-x^m=a^m\left(\mathbf{1}-\frac{x^m}{a^m}\right);$$

$$\sqrt[3]{(a^3-x^3)}=a\sqrt[3]{\left(\mathbf{1}-\frac{x^3}{a^3}\right)};$$

$$\sqrt{a^2+x^2}=a\sqrt{\left(\mathbf{1}+\frac{x^2}{a^2}\right)};$$

$$(\sqrt{a}+\sqrt{b})\,(\sqrt{a}-\sqrt{b})=a-b\ ;\ \text{whence we have:}$$

$$\frac{a-b}{\sqrt{a}-\sqrt{b}}=\sqrt{a}+\sqrt{b}\ ;\ \text{also}$$

$$\frac{a-b}{\sqrt{a}+\sqrt{b}}=\sqrt{a}-\sqrt{b}\ ;$$

$$\frac{x^2+2x-3}{x^2+5x+6}=\frac{x^2+2x+1-4}{x^2+4x+4+x+2}=\frac{(x+1)^2-4}{(x+2)^2+x+2}\ ;$$

$$\frac{x^2+2x-3}{x^2+5x+6}=\frac{(x+1+2)\,(x+1-2)}{(x+2)\,(x+2+1)}=\frac{x-1}{x+2}.$$

216.

In all calculations with powers and roots, most careful attention must be given to the signs.

The conditions governing these may be reduced to five rules which we have intentionally postponed until the end of this book, and which are here now given:

(1) Every power of a positive quantity is positive, thus:

$$(+a)^n=+a^n$$

For a negative quantity all the *even* powers are positive, and and all the *odd* powers are negative. This follows because the product of an even number of minus factors is plus, and the product of an uneven number of minus factors is minus, thus:

$$(-3)^2=9 \text{ for } (-3)^2=-3\cdot-3=+9$$

$$(-3)^3=-3\cdot-3\cdot-3=9.-3=-27$$

$$(-3)^4=-3\cdot-3-3\cdot-3=9.9=81$$

If n is any whole number, then $2n$ is always even, and $2n+1$ always odd, and we have in general:

$$(-a)^{2n}=+a^{2n}; \quad \text{and} \quad (-a)^{2n+1}=-a^{2n+1}$$

EXAMPLES:

$$(-2)^3=-8; \qquad (-2)^4=16; \qquad (-\tfrac{1}{2})^3=\tfrac{1}{8};$$
$$(-a)^{10}=+a^{10}; \qquad (-a)^{11}=-a^{11}; \qquad (-4)^2-16.$$

We must distinguish between $-a^2$ and $(-a)^2=a^2$; $-a^2$ is read "minus a square," $(-a)^2$ is read, "minus a *squared*," or "the square of minus a." In the same manner $\tfrac{1}{2}a^2$ is read "one-half a square," while $(\tfrac{1}{2}a)^2$ is read "the square of one-half a," and is $=\tfrac{1}{4}a^2$.

(3) Conversely it follows that every odd root of a positive quantity must be positive, while every odd root of a negative quantity must be negative, thus:

$$\sqrt[3]{(+8)}=+2 \quad \sqrt[3]{(-8)}=-2$$
$$\sqrt[3]{(-27)}=-3 \text{ for } +3^3+27, \text{ while } (-3)^3=-27$$

There is no quantity but $+2$ which can give $+8$ when cubed, and -2 is the only quantity which when cubed gives -8.

In general we have:

$$\sqrt[2n+1]{(+a)}=+\sqrt[2n+1]{a} \text{ and :}$$
$$\sqrt[2n+1]{(-a)}=-\sqrt[2n+1]{a}$$

(4) Every *even* root of a positive quantity may be either positive or negative, since the *even* powers of both negative and positive quantities are positive.

For example:

$$(+3)^2=(-3)^2=+9$$

and therefore $\sqrt{9}=\pm3$, (read "plus or minus three").

Likewise $\sqrt{4}=\pm2$; $\sqrt{\tfrac{4}{9}}=\pm\tfrac{2}{3}$; $\sqrt{16}=\pm4$; $\sqrt[4]{81}=\pm3$.

In general we have:

$$\sqrt[2n]{(+a)}=\pm\sqrt[2n]{a}$$

In the preceding examples, and in many of the following, the double sign $(\pm)$ has, for sake of simplicity, not been used. In practice, however, the double sign should always be prefixed when an even root has been extracted, unless there is some other means of determining whether the root is negative or positive. Sometimes it is possible to know this from given conditions, thus:

$$\sqrt{-a}\cdot{-a}=\sqrt{a^2}=\sqrt{(-a)^2}=-a$$

(5) Finally we have the case in which the radical sign with an *even* exponent stands over a *negative* quantity, as for example:

$$\sqrt{-4}\; ;\; \sqrt[4]{-16}\; ;\; \sqrt{-a}\; ;\; \text{or since each root has the double sign,}$$

$$\pm\sqrt{-4}\; ;\; \pm\sqrt{-a}\; ;\; \text{etc.}$$

But since there is no number, either plus or minus, which when raised to an *even* power can have a minus sign, so it follows that no real root can be extracted from a negative quantity. Such roots can therefore only be indicated, thus,

$$\sqrt{-4}=\sqrt{-4}\; ;\; \sqrt[4]{-5}=\sqrt[4]{-5}\; ;\; \sqrt[6]{-a}=\sqrt[6]{-a}\; ;$$

for we can conceive of no quantity which when raised to the 2nd, 4th, or 6th powers would give -4, -5, or $-a$.

Such expressions therefore as $\sqrt{-a}$, $\sqrt{-a^2}$, etc., are called *imaginary* quantities. See appendix, § 325, 326.

BOOK XVII.

The Solution of Quadratic Equations.

217.

DEFINITIONS. In any equation, after the parentheses have been removed, and both sides cleared of fractions, if the unknown quantity does not appear in any higher power than the first degree, the equation is called a *simple* equation, or an equation of the *first* degree. When the unknown appears in any higher power than the first, we have an equation of a higher degree, and the degree of an equation, is the same as that of the exponent of the *highest power* of the unknown which it contains.

Higher equations are *pure* or *mixed*, according as they contain the unknown in but one degree or in several degrees. We have therefore the following classification, with examples :

$$\left.\begin{array}{l} 2x=6 \\ 3x-7=14-x \end{array}\right\} \text{Equations of the First Degree or Simple Equations.}$$

$$\left.\begin{array}{l} x^2=9 \\ 2x^2+16=\tfrac{8}{3}x^2 \end{array}\right\} \text{Pure Equations of the Second Degree or Pure Quadratic Equations.}$$

$$\left.\begin{array}{l} x^2+2x=16 \\ 8-\tfrac{3}{4}x=6x^2 \end{array}\right\} \text{Mixed Quadratic Equations, containing the unknown quantity in the first as well as in the second power.}$$

$$\left.\begin{array}{l} x^3=27 \\ x^3+1\tfrac{36}{31}-7x^3=-\tfrac{2}{3}x^3 \end{array}\right\} \text{Pure Equations of the Third Degree.}$$

In general :

$$\left.\begin{array}{l} x^n=a \\ ax^n+bx^n=c-dx^n \end{array}\right\} \text{Pure Equations of the } n \text{th Degree.}$$

$$x^n+ax^{n-1}=x+b \qquad \text{Mixed Equations of the } n \text{th Degree.}$$

The general theory of equations of the higher degree belongs to the subject of Higher Algebra or analysis. In elementary algebra only pure equations and mixed quadratic equations are discussed.

218.

The solution of pure quadratic equations offers no difficulty. It is only necessary to clear the unknown quantities (x^2) from parentheses and fractions, collect them by the sign $+$ on one side of the equation and then extract the square root of both sides.

Reducing the pure quadratic equation to the general form:

$$x^2 = q$$

in which x is the unknown, and q the known or given quantity, and we have, by extracting the root:

$$\sqrt{x^2} = \sqrt{q}$$
$$\text{or} \quad x = \pm\sqrt{q} \qquad (\S 216, 4)$$

Any value of the unknown, which, when substituted in the original equation, satisfies its conditions, is called a *solution* or *root* of that equation (§ 100). The pure quadratic equation has, therefore, two roots, of equal value but opposite signs, namely: $x = +\sqrt{q}$ and $x = -\sqrt{q}$.

EXAMPLE 1. What are the values for x in the following equation?

$$2x^2 - 3 = 69$$

Solution :
$$2x^2 = 72$$
$$x^2 = 36$$
$$x = \sqrt{36} = \pm 6$$

The equation is satisfied equally well by $+6$ or -6

EXAMPLE 2. Find x, from the following equation:

$$\frac{x^2}{4} + 7 - \frac{2x^2}{3} = \frac{5x^2}{6} - 153.$$

Solution. First collecting all the members containing x^2 to the left hand side, we have :

$$\frac{x^2}{4} - \frac{2x^2}{3} - \frac{5x^2}{6} = -160$$

Multiplying by the common denominator 12, to clear of fractions :

$$3x^2 - 8x^2 - 10x^2 = -12 \cdot 160$$

Collecting the coefficients of x^2 :

$$-15x^2=-12\cdot160$$

$$x^2=\frac{12\cdot160}{15}=128$$

$$x=\pm\sqrt{128}$$

Whence $x=+11.314$ or -11.314

EXAMPLE 3. Required the value of x in terms of the quantities a, b, c, h, m, and n,. The relation of these quantities is given in the following equation.

$$\frac{cx^2}{m}+n-\frac{bx^2}{n}=-\frac{ax^2}{n}+\frac{ab}{c}-\frac{cx}{m}$$

Solution. Separating the terms containing the unknown from the known, we have :

$$\frac{ax^2}{n}-\frac{bx^2}{n}+\frac{cx^2}{m}+\frac{cx^2}{m}=\frac{ab}{c}-n$$

Now multiplying by mn, the common denominator of the unknown terms, we have :

$$amx^2-bmx^2+2cnx^2=mn\frac{(ab-ch)}{c}$$

or $\quad (am-bm+2cn)x^2=mn\frac{(ab-ch)}{c}$

$$x^2=\frac{mn(ab-ch)}{c(am-bm+2cn)}$$

whence $x=\pm\sqrt{\dfrac{mn(ab-ch)}{c(am-bm+2cn)}}$

EXAMPLE 4. Find a number such that when 4 is added to it, and the sum divided by 3, the result shall be equal to 3 divided by the required number minus 4.

Solution. If we call the required number x, we have from the given conditions, the equation :

$$\frac{x+4}{3}=\frac{3}{x-4}$$

Multiplying by the common denominator $3(x-4)$ we have :

$$(x+4)(x-4)=9$$
$$x^2-16=9$$
$$x^2=25$$
$$x=\pm5$$

If we take the $+$ sign and substitute in the original equation we have:

$$\frac{5+4}{3}=\frac{3}{5-4}=3$$

If we take the $-$ sign, we have:

$$\frac{-5+4}{3}=\frac{3}{-5-4}=-\tfrac{1}{3}$$

219.

Any pure equation of the nth degree may be solved in a manner similar to that above given for pure quadratic equations.

It is only necessary to collect the nth powers of the unknown quantities on one side, and all the known quantities on the other side, and then extract the nth root of both sides; which can easily be done by means of a table of logarithms, as will be explained in §234.

Since every pure equation of the nth degree has n roots, and hence n different values for x, which will satisfy it, it is ordinarily the custom to give only the real roots in elementary algebra. The other roots which give imaginary values can only be found by means of higher mathematics.*

We can readily find the real value of x from the following equation, for example:

$$x^3+\frac{136}{81}-7x^3=-\tfrac{1}{3}x^3$$

Transposing:

$$x^3-7x^3+\tfrac{1}{3}x^3=-\frac{136}{81}$$

Collecting:

$$-5\tfrac{2}{3}x^3=-\frac{136}{81}$$

$$\frac{17x^3}{3}=\frac{136}{81}$$

$$x^3=\frac{3\cdot 136}{81\cdot 17}=\frac{8}{27}$$

$$x=\sqrt[3]{\tfrac{8}{27}}=\tfrac{2}{3}$$

*The equation $x^4-4x^3-x^2-16x=12$, for example, has for roots; 1, 2, -2, and 3. The equation $x^3=8$, has three roots; 2, $-1+\sqrt{-3}$, and $-\sqrt{}+3$. The equation x^4-4 has four roots: $\sqrt{}2$, $-\sqrt{}2$, $\sqrt{}-2$ and $-\sqrt{}-2$.

From the equation:

$$x^{30}+30x^{30}=10$$

we have, $\quad x^{30}=\dfrac{10}{31}$

whence, $\quad x=\sqrt[30]{(\tfrac{10}{31})}$ etc.

(See §284)

Mixed Quadratic Equations.

220.

The solution of equations of this sort, (first discovered by an Arabian mathematician, Mohammed-Ben-Musa) involves the exercise of some judgment.

All members containing the square of the unknown quantity must be collected into one member, as must also all members containing the first power of the unknown, and the equation so arranged that has only three members. The square of the unknown quantity without denominators or coefficients, and preceded by a plus sign, is followed by the first power of the unknown with its coefficient, forming one side of the equation; while the known quantities are collected on the other side. The equation will then have the form:

$$x^2+px=q$$

in which p and q are known, and may be either positive or negative, fractional or whole numbers.

For example, the following equation

$$\frac{2x}{3}+10\tfrac{1}{4}+\tfrac{3}{4}x^2=2x^2+10-3x$$

may be brought into the required form as follows:

$$\frac{3x^2}{4}-2x^2+\frac{2x}{3}+3x=-\tfrac{1}{4}$$

Multiplying both sides by 12:

$$9x^2-24x^2+8x+36x=-\tfrac{1}{4}\cdot12$$

$$-15x^2+44x=-3$$

$$15x2-44x=3$$

$$x^2-\frac{44}{15}x=\frac{3}{15}$$

In a similiar manner the following equation may be reduced :

$$\frac{cx}{n}+c-\frac{bx^2}{m}\frac{hx}{c}=\frac{hx}{c}-\frac{ax^2}{n}+d$$

collecting :
$$\frac{ax^2}{n}-\frac{bx^2}{m}+\frac{cx}{n}-\frac{hx}{c}=d-c$$

Clearing of fractions :

$$macx^2-nbcx^2+mc^2x-mnhx=mnc(d-c)$$

$$(mac-nbc)x^2+m(c^2-nh)x=mnc(d-c)$$

$$x^2+\frac{m(c^2-nh)}{c(ma-nb)}\cdot x=\frac{mnc(d-c)}{c(ma-nb)}$$

221.

After the mixed quadratic equation has been brought to the form :

$$x^2+px=q \qquad (1)$$

as shown in the preceding section, the two members on the left side must be considered as a portion of a square, i. e., as

$$x^2+2\cdot\tfrac{1}{2}px=q$$

We see now by inspection that the quantity which is lacking to make the left side a perfect square is the square of half the coefficient of x (see § 213), that is $(\tfrac{p}{2})^2$, and we may add this to both sides of the equation without changing its value.

We then have from equation (1) :

$$x^2+px+\left(\frac{p}{2}\right)^2=\frac{p^2}{4}+q \qquad (2)$$

This now has for its left side a complete square, of which the root is $x+\tfrac{p}{2}$. We have now only to extract the square root of both sides of the equation, indicating the operation for the right hand side (see § 214) and we have an equation of the first degree. Thus from (2), we have :

$$x+\frac{p}{2}=\pm\surd\left(\frac{p^2}{4}+q\right) \qquad \text{(see § 216,4)}$$

whence :
$$x=-\frac{p}{2}\pm\surd\left(\frac{p^2}{4}+q\right)$$

or bringing the quantities under the radical to the same denominator, and extracting the root of the denominator :

$$x=-\frac{p}{2}\pm\surd\left(\frac{p^2+4q}{4}\right)=\frac{-p\pm\surd\overline{p^2+4q}}{2}$$

In order to make the above method of "completing the square," as it is called, perfectly clear to the beginner, it must be again stated that the square of a binomial quantity is equal to the square of the first member, plus twice the product of the first by the second member, plus the square of the second member, or in a formula :

$$(x+a)^2 = x^2 + 2ax + a^2.$$

Now if we only had $x^2 + 2ax$, and wished to find out what must be added to make the expression a complete square, we see that the coefficient of x in the second term, i. e., $2a$, is double the second term of the required root $x+a$, and hence, if we take one-half the coefficient of x, i. e., one-half of $2a = a$, and square it, and add it to the imperfect square $x^2 + 2ax$, we shall have the complete square $x^2 + 2ax + a^2$. As a^2 added to *both* sides of the equation does not change the truth of the equality, we have a perfect right to do it, and as it does not introduce any unknown quantity on the right hand side, the solution of the equation is not hindered.

222.

EXAMPLE 1. What values of x satisfy the following equation ?

$$5x^2 = 30x - 40 \qquad (1)$$

Solution: First arranging the members according to §220, we have:

$$5x^2 - 30x = -40$$
$$x^2 - 6x = -8 \qquad (2)$$

The coefficient of x is 6, and one-half this squared and added on both sides will complete the square, thus :

$$x^2 - 6x + 9 = 9 - 8$$
$$x^2 - 6x + 9 = 1 \qquad (3)$$

Extracting the square root of both sides:

$$x - 3 = \pm 1 \qquad (4)$$
$$x = 3 \pm 1$$

The values of x are therefore $3 + 1 = 4$ and $3 - 1 = 2$, either of which will satisfy equation (1). It will also be seen that either of these values when substituted in equation (4) will give, when squared, equation (3).

223.

EXAMPLE 2.   Find the values of x from the following equation :

$$\frac{2x}{3}+10\tfrac{1}{4}+\frac{3x^2}{4}=2x^2+10-3x \qquad (1)$$

Solution: Arranging the numbers according to §220, we have :

$$x^2-\frac{44}{15}x=\frac{3}{15} \qquad (2)$$

Adding $(\tfrac{22}{15})^2$ to both sides, to complete the square, we have :

$$x^2-\frac{44}{15}x+\left(\frac{22}{15}\right)^2=\frac{22^2}{15^2}+\frac{3}{15} \qquad (3)$$

Extracting the square root :

$$x-\frac{22}{15}=\pm\sqrt{\left(\frac{22^2}{15^2}+\frac{3}{15}\right)} \qquad (4)$$

In order to extract the square root of the right hand side of equation (4) we first collect into one member over a common denominator (§214). We therefore have :

$$\frac{22^2}{15^2}+\frac{3}{15}=\frac{22}{15^2}+\frac{3\cdot15}{15^2}=\frac{484+45}{15^2}=\frac{529}{15^2}$$

Whence :

$$x-\frac{22}{15}=\pm\sqrt{\frac{529}{15^2}}$$

$$x=\frac{22}{15}\pm\frac{\sqrt{529}}{15}$$

$$x=\frac{22\pm\sqrt{529}}{15}$$

$$x=\frac{22\pm23}{15}$$

whence :

$$x=\frac{22+23}{15}=3$$

and :

$$x=\frac{22-23}{15}=-1\tfrac{1}{15}$$

224.

EXAMPLE 3.   Reduce the following equation to the values of x:

$$acx-bcx=ab-c^2x^2 \qquad (1)$$

Arranging in proper order :

$$c^2x^2+c(a-b)x=ab$$

$$x^2+\frac{a-b}{c}x=\frac{ab}{c^2}$$

Completing the square :

$$x^2+\frac{a-b}{c}x+\left(\frac{a-b}{2c}\right)^2=\left(\frac{a-b}{2c}\right)^2+\frac{ab}{c^2}$$

Extracting the root :

$$x+\frac{a-b}{2c}=\pm\sqrt{\frac{(a-b)^2+4ab}{4c^2}}$$

$$=\pm\frac{\sqrt{(a-b)^2+4ab}}{2c}$$

Now taking the quantity under the radical sign and removing the parentheses, we have :

$$(a-b)^2+4ab=a^2-2ab+b^2+4ab$$

$$=a^2+2ab+b^2$$

$$=(a+b)^2$$

or $\quad x+\dfrac{a-b}{2c}=\pm\dfrac{\sqrt{(a+b)^2}}{2c}$

$$=\pm\frac{a+b}{2c}$$

whence :

$$x=-\frac{a-b}{2c}\pm\frac{a+b}{2c}=\frac{b-a\pm(a+b)}{2c}$$

If we take the upper sign, the value of $x=\dfrac{b}{c}$; of the lower sign, $x=-\dfrac{a}{c}$ either of which values will be found to satisfy equation (1).

· 225.

EXAMPLE 4. The sum of \$16000 is to be divided equally among a certain number of persons. If there were two less persons, each share would be \$4000 greater; how many persons were there?

Solution. Let x be the number of persons, then the share of each will be $\dfrac{16000}{x}$

If now there were two less persons the share of each would be $\dfrac{16000}{x-2}$, and since each share would then be $4000 greater, we have:

$$\frac{16000}{x-2}=\frac{16000}{x}+4000$$

Dividing both sides by 4000 we have:

$$\frac{4}{x-2}=\frac{4}{x}+1$$

Multiplying by $x(x-2)$ to clear of fractions:

$$4x=4x-8+x^2-2x$$
$$x^2-2x=8$$
$$x^2-2x+1=9$$
$$x-1=\pm3$$
$$x=4$$

We must here take the upper sign since $x=1-3=-2$ gives a negative value, which while it satisfies the equation, is an impossible solution of a problem requiring a positive number of people. This, then, is a question in which the correct sign is determined by outside considerations. (See §144).

226.

EXAMPLE 5. A lady being asked her age, said 53 times my age exceeds the number 696, by the square of my age. How old was she?

Solution. Let $x =$ the age, then:

$$53x=696+x^2$$
$$x^2-53x=-696$$
$$x^2-53x+\left(\frac{53}{2}\right)^2=\frac{53^2}{4}-696$$
$$x-\frac{53}{2}=\pm\frac{\sqrt{53^2-696\cdot4}}{2}=\frac{\pm\sqrt{25}}{2}$$
$$x=\frac{53\pm5}{2}$$

whence:

$$x=\frac{53-5}{2}=24$$

or: $\quad =\dfrac{53+5}{2}=29$

As a matter of courtesy, of course the lower value would be chosen.

Beginners often wonder at the fact that two answers can be obtained to one equation and think it a defect in the method. It is, however, no defect, both results being strictly correct, and as will be seen in the applications of algebra to geometry, giving relations which are identical in meaning for all possible cases.

227.

EXAMPLE 6. A line, 10 inches long, is to be divided into two such parts that the smaller part shall bear the same proportion to the larger part, that the larger part bears to the whole length. What are the two parts?

Solution. Let x be the smaller part, and $a-x$ the greater part : a, being the length of the whole line. We then have $\dfrac{a-x}{x}$ must give the same quotient as $\dfrac{a}{a-x}$, and these two expressions must therefore be equal to each other.

$$\frac{a-x}{x} = \frac{a}{a-x}$$

whence :
$$(a-x)(a-x) = ax$$
$$a^2 - 2ax + x^2 = ax$$
$$x^2 - 2ax - ax = -a^2$$
$$x^2 - 3ax = -a^2$$

$$x^2 - 3ax + \left(-\frac{3a}{2}\right)^2 = \frac{9a^2}{4} - a^2$$

$$\left(x - \frac{3a}{2}\right)^2 = \frac{5a^2}{4}$$

$$x - \frac{3a^2}{2} = \frac{\sqrt{5a^2}}{2}$$

$$x = \frac{3a \pm a\sqrt{5}}{2}$$

whence :
$$x = a\left(\frac{3 \pm \sqrt{5}}{2}\right) = a\left(\frac{3 \pm 2.236}{2}\right)$$

whence :
$$x = 0.382a$$

The smaller part is $x = 0.382a$ and the greater part :
$$a - x = a - 0.382a = 0.618a$$
and since a equals 10 inches, $x = 3.82$ ins., and $a - x = 6.18$ ins.

Here we must use the lower sign (—), for we see that the use of the + sign for the root would make the desired part x, greater than the whole line, and hence the greater part a—x would come out negative. This is only impracticable, however, because we are considering a *line*, and if the problem were to divide the *number* 10 into two such parts, both solutions would be equally practicable. We may notice also that the value of x is irrational and can hence only be expressed approximately.

228.

EXAMPLE 7. Required the values for x which shall satisfy the following equation :

$$x^2 = 2x - 5 \qquad (1)$$

Solution.
$$x^2 - 2x = -5$$
$$x^2 - 2x + 1 = -4$$
$$x - 1 = \pm\sqrt{-4}$$
$$x = 1 \pm\sqrt{-4}$$

Here the root comes out an imaginary quantity, as it involves the square root of a negative quantity (see § 325), but it is none the less correct, as we see by substituting in the original equation:

$$(1 \pm\sqrt{-4})^2 = 2(1 \pm\sqrt{-4}) - 5$$

Removing the parentheses :

$$1 \pm 2\sqrt{(-4)} - 4 = 2 \pm 2\sqrt{(-4)} - 5$$
$$-3 \pm 2\sqrt{-4} = 3 \pm 2\sqrt{-4}$$

229.

EXAMPLE 8. Determine the value of x in terms of a, b and c, in the following equation :

$$\frac{bc}{a} - ax - c = \frac{bx^2}{a} - \frac{ax^2}{b} - bx$$

Solution. We have :

$$\frac{ax^2}{b} - \frac{bx^2}{a} - ax + bx = c - \frac{bc}{a}$$

Multiplying by ab :

$$a^2x^2 - b^2x^2 - a^2bx + ab^2x = abc - b^2c$$
$$(a^2 - b^2)x^2 - ab(a - b)x = bc(a - b)$$
$$x^2 - \frac{ab(a-b)}{a^2-b^2}x = \frac{bc(a-b)}{a^2-b^2}$$
$$x^2 - \frac{ab}{a+b}x = \frac{bc}{a+b} \qquad \text{(see § 91)}$$

$$x^2 - \frac{ab}{a+b}x + \left[\frac{ab}{2(a+b)}\right]^2 = \frac{a^2b^2}{4\,(a+b)^2} + \frac{bc}{a+b}$$

$$x - \frac{ab}{2(a+b)} = \frac{\sqrt{a^2b^2 + 4bc(a+b)}}{2(a+b)}$$

$$x = \frac{ab \pm \sqrt{a^2b^2 + 4bc\,(a+b)}}{2(a+b)}$$

230.

When the unknown quantity appears in an equation with a fractional exponent or under a radical sign, the equation is called an irrational equation. An irrational equation may, however, be made rational by collecting all the radicals on one side and then raising both sides to the power indicated by the exponent of the root. This follows, because, of course, the value of the unknown or the truth of the equation, is not affected by the squaring, cubing, etc., of *both sides* of the equation.

Thus with the equation :

$$ax = b \pm \sqrt{x}$$

$$ax - b = \pm \sqrt{x}$$

Now squaring both sides, the radical sign disappears, because the square of $\pm\sqrt{x}$ is equal to x, and we have the rational equation :

$$a^2x^2 - 2abx + b^2 = x$$

whence :

$$a^2 - x^2 - 2abx - x = -b^2$$

$$x^2 - \frac{(1+2ab)}{a^2}x = -\frac{b^2}{a^2}$$

$$x \frac{-1+2ab}{2a^2} = \pm\sqrt{\left[\frac{(1+2ab)^2}{4a^4} - \frac{b^2}{a^2}\right]}$$

$$x = \frac{1+2ab \pm \sqrt{(1+2ab)^2 - 4a^2b}}{2a^2}$$

$$x = \frac{1+2ab \pm \sqrt{1+4ab}}{2a^2}$$

If an equation has several irrational members, the foregoing operation must be repeated until they are one after the other made rational : as in the following equation :

$$\sqrt{2x+7} = 2 + \sqrt{5-4x}$$

Here we square both sides, and observing that in general $(\sqrt{a+b})^2=a+b$; also that $(a+\sqrt{b})^2=a^2+b+2a\sqrt{b}$; and also that

$$[a\sqrt{b-c}]^2=a^2(b-c)$$

We have :

$$2x+7=4+5-4x+4\sqrt{5-4x}$$
$$6x-2=4\sqrt{5-4x}$$

Now squaring again :

$$36x^2-24x+4=80-64x$$
$$36x+40x=76$$
$$x^2+\tfrac{10}{9}x=\tfrac{19}{9}$$
$$x^2+\tfrac{10}{9}x+(\tfrac{5}{9})^2=\frac{25}{9^2}+\frac{19}{9}=\frac{25+9\cdot19}{9^2}$$
$$x+\tfrac{5}{9}=\pm\frac{\sqrt{196}}{9}$$
$$x=\frac{-5\pm14}{9}$$

whence the values of x are :

$$x=1 \text{ and } x=-\tfrac{19}{9}$$

231.

Equations of any higher degree, when they can be brought into the form

$$x^{2m}+px^m=q$$

can be solved in the same manner as quadratic equations. That is, when only two powers of the same unknown quantity appear, and the greater exponent is twice the lesser exponent, the equation can readily be converted into a true quadratic equation.

Thus, suppose we call $x^m=z$, then $x^{2m}=(x^m)^2=z^2$, and substituting these in the equation :

$$x^{2m}+px^m=q$$

we get :

$$z^2+pz=q$$

whence :

$$z=\frac{-p\pm\sqrt{p^2+4q}}{2}$$

Then substituting for z, its value x^m, we get :

$$x^m=\frac{-p\pm\sqrt{p^2+4q}}{2}$$
$$x=\sqrt[m]{\left(\frac{-p\pm\sqrt{p^2+4q}}{2}\right)}$$

If the mth root is an *even* root the $\pm$ sign must also be placed before it.

232.

EXAMPLE. Divide the number 18 into two such factors that when each factor is squared, their sum shall be equal to 45.

Solution. Let x be one factor, then $\dfrac{18}{x}$ will be the other, and

$$x^2 + \frac{18^2}{x^2} = 45$$

$$x^4 + 18^2 = 45x^2$$

$$x^4 - 45x^2 = -324$$

Now putting $x^2 = z$, and $x^4 = z^2$, we have :

$$z^2 - 45z = -324$$

$$z^2 - 45z + (\tfrac{45}{2})^2 = \frac{45^2 - 4 \cdot 324}{4}$$

$$z - \tfrac{45}{2} = \frac{\sqrt{729}}{2}$$

$$x^2 = \frac{45 \pm 27}{2}$$

$$x = \pm \sqrt{\left(\frac{45 \pm 27}{2}\right)}$$

Since we may take the two like or two unlike signs, there are *four* values possible for x, and these come in pairs of equal magnitude and opposite sign, as follows :

for $++$ $x = 6$ and $\dfrac{18}{x} = 3$

for $+-$ $x = 3$ and $\dfrac{18}{x} = 6$

for $-+$ $x = -6$ and $\dfrac{18}{x} = -3$

for $--$ $x = -3$ and $\dfrac{18}{x} = -6$

233.

EXAMPLE. Divide the number 12 into two such factors that the difference of their cubes $= 37$.

Solution. Let x be one factor, and $\dfrac{12}{x}$ the other, and we have :

$$x^3 - \frac{12^3}{x^3} = 37$$

$$x^6 - 12^3 = 37x^3$$

$$x^6 - 37x^3 = 12^3$$

Putting $x^3 = z$, and $x^6 = x^3 \cdot x^3 = z^2$, we have ;

$$z^2 - 37z = 1728$$

$$z = \frac{37 \pm \sqrt{37^2 + 4 \cdot 1728}}{2}$$

whence, $\quad x^3 = \dfrac{37 \pm \sqrt{8281}}{2}$

whence, $\quad x = \sqrt[3]{64} = 4$, or $x = \sqrt[3]{-27} = -3$

and $\quad \dfrac{12}{x} = \dfrac{12}{4} = 3$, or $\dfrac{12}{x} = \dfrac{12}{-3} = -4$

EXAMPLES :

(1) $\quad \dfrac{50}{2x+1} + 3 = \dfrac{8x-3}{9-4x}$ $\qquad$ Ans. $\quad x = 2 \pm 4$

(2) $\quad b = \dfrac{a - \sqrt{a^2 - x^2}}{a + \sqrt{a^2 - x^2}}$ $\qquad$ Ans. $\quad x = \pm \dfrac{2a\sqrt{b}}{1+b}$

(3) $\quad \sqrt[2m]{2x^2 + 4ax - b^2} = \sqrt[m]{a+x}$ $\quad$ Ans. $x = \pm -a \pm \sqrt{2a^2 + b^2}$

NOTE.—In the above example (3) both sides should be raised to the $2m$th power (see § 207).

(4) $\quad 2x\sqrt[3]{x} - \dfrac{3x}{\sqrt[3]{x}} = 20$ $\qquad$ Ans. $\quad x = \pm 8$ and $\pm \sqrt{\dfrac{-125}{8}}$

In example (4) it should be written $2x^{\frac{4}{3}} - 3x^{\frac{2}{3}} = 20$, and then place $x^{\frac{2}{3}} = z$, etc., as in § 231.

Quadratic Equations With Several Unknown Quantities.

234.

If, we have n equations, and n unknown quantities, and some or all of the equations are quadratic, each unknown must be determined by elimination.

When there are more than two equations, however, the solution is only possible in certain favorable cases.

235.

EXAMPLE 1. Suppose we have the sum of two quantities, x and y, is equal to s, and their product $xy=p$; s and p being known quantities, such as 10 and 24, how can the two unknowns, x and y, be determined from s and p?

Solution. We have :

$$x+y=s \qquad (1)$$

$$xy=p \qquad (2)$$

Multiplying the first equation by x, we have :

$$x^2+xy=sx \qquad (3)$$

Subtracting the second equation :

$$x^2=sx-p$$

whence, $\quad x^2-sx=-p$

$$x=\frac{s\pm\sqrt{s^2-4p}}{2} \qquad (4)$$

which substituted in (1) gives :

$$y=\frac{s\mp\sqrt{s^2-4p}}{2} \qquad (5)$$

The inverted sign ($\mp$) in (5) means that when we take the upper sign for x we must take the lower for y, and *vice versa*.

236.

EXAMPLE 2. Given the sum of two quantities and the sum of their squares, required to determine the quantities. We have:

$$x+y=a \qquad (1)$$

$$x^2+y^2=b \qquad (2)$$

Solution. The shortest way in this case is to square the first equation, and subtract the second from it. This gives :

$$2xy=a^2-b \qquad (2)$$

Then subtracting (3) from (2), we have :

$$x^2-2xy+y^2=2b-a^2$$

$$\text{and} \quad x-y=\sqrt{2b-a^2} \qquad (4)$$

From (4) and (1) we then have : $\qquad$ (§ 167)

$$x=\frac{a\pm\sqrt{2b-a^2}}{2}$$

$$y=\frac{a\mp\sqrt{2b-a^2}}{2}$$

237.

EXAMPLE 3. Let the product of two quantities equal p, and the sum of their squares $=a$. Required the quantities in terms of a and p.

We have:

$$xy=p \qquad (1)$$
$$x^2+y^2=a \qquad (2)$$

Solution. If we multiply (1) by 2 we have $2xy=2p$. If now we add this to the equation (2) we have:

$$x^2+2xy+y^2=a+2p$$

and if we subtract it, we have:

$$x^2-2xy+y^2=a-2p$$

whence, $\quad x+y=\sqrt{a+2p} \qquad (3)$

and, $\quad x-y=\sqrt{a-2p} \qquad (4)$

whence, $\quad x=\dfrac{\pm\sqrt{a+2p}\pm\sqrt{a-2p}}{2} \qquad (5)$

and, $\quad y=\dfrac{\pm\sqrt{a+2p}\mp\sqrt{a-2p}}{2} \qquad (6)$

(See § 167)

Squaring equations (5) and (6): (See § 186)

$$x^2=\frac{a+2p+a-2p\pm2\sqrt{a^2-4p^2}}{4}=\frac{a\pm\sqrt{a^2-4p^2}}{2}$$

whence, $\quad x=\pm\sqrt{\left\{\dfrac{a\pm\sqrt{a^2-4p}}{2}\right\}}$

and, $\quad y=\pm\sqrt{\left\{\dfrac{a\mp\sqrt{a^2-4p}}{2}\right\}}$

238.

EXAMPLE 4. Given:

$$3x+2y=8 \qquad (1)$$
$$4x^2-3y^2=13 \qquad (2)$$

Solution. Find the value of x in (1) and substitute in (2), and we get:

$$y=1, \quad x=2$$

or $\quad y=-\tfrac{13}{11}$ and $x=\tfrac{22}{11}$

239.

EXAMPLE. Given :

$$x^2+xy+y^2=7 \qquad (1)$$
$$y^2+yz+z^2=19 \qquad (2)$$
$$x^2+xz+z^2=13 \qquad (3)$$

Solution. Subtracting (2) from (1) and (2) from (3), we get :

$$x+y+z=\frac{-12}{x-z}$$

and $x+y+z=\dfrac{-6}{x-y}$

Equating these two values, we have :

$$\frac{12}{x-z}=\frac{6}{x-y}$$

or, $x=2y-z$

Substituting this value of x in (3) and subtracting the result from (2) gives :

$$z=\frac{y^2+2}{y}$$

Substituting this value of z in (2), gives :

$$y^4-\tfrac{13}{3}y^2=-\tfrac{4}{3}$$

whence, according to § 231, we get :

$$y=\pm 2, \quad =\pm\sqrt{\tfrac{1}{3}}$$

$$z=\pm 3, \quad =\pm\sqrt{\tfrac{7}{3}}$$

$$x=\pm 1, \quad =\mp\sqrt{\tfrac{5}{3}}$$

BOOK XVIII.

Series.

1. Arithmetical Series.

240.

Any series of numbers in which we find the *difference* obtained by subtracting any of its members from the next following member to be a *constant* quantity, is called an Arithmetical Series, or Arithmetical Progression. Such a series is called an *increasing* or *diminishing* series, according as the successive members become uniformly greater or smaller. Examples of such Arithmetical Progression are :

$$0, 1, 2, 3, 4, 5. 3 \ldots \quad (1)$$
$$5, 6, 7, 8, 9. 10 \ldots \quad (2)$$
$$4, 9, 14, 19, 24, 29 \ldots \quad (3)$$
$$8, 5, 2, -1, -4, -7 \quad (4)$$

In the first and second series the constant difference is 1, in the third it is 5, in the fourth, diminishing series it is—3.

241.

If we have the first member of an arithmetical series given, and also the constant difference, we can extend the series to any desired extent; the difference added to the second member will give the third, or, what is the same thing, twice the difference added to the first member will give the third, etc., so that to obtain the hundredth member it is just the same if we add the difference to the 99th member, or add 99 times the difference to the first member.

If, for example, the first member be 3 and the difference =4, we have:

(1)	(2)	(3)	(4)	10th Member
3	3+4	3+2·4	3+3·4 . . .	3+9·4
or, 3	7	11	15 . . .	39

If 15 is the first member and —5 the difference :

(1)	(2)	(3)	(4)	20th Member
15	15—5	15—2·5	15—3·5 . . .	15—19·5
or, 15	10	5	0 . ·.	—80

242.

In discussing arithmetical series the following five magnitudes, and their corresponding symbols are used, namely : The first member $=a$, the difference $=d$, the last member l (*terminus*), the number of members $=n$ (*numerus*), and finally the sum $=s$, of all the members added together. Each of these five magnitudes, a, d, l, n and s, is a determinate function of any three of the others, and if the latter are given in numerical values, the former can be readily determined from them, without any necessity for constructing the whole series. The most important problems in this connection are the determination of the magnitude of any particular member, and the determination of the sum of the whole series.

243.

PROBLEM. Let it be required to find a general formula by which the magnitude of any particular member of the series may be determined, when its number $=n$, (i. e., its place in the series) the first member $=a$, and the difference $=d$, are given.

Solution. If the first member $=a$, the second member $= a+d$, the third $=a+2d$ etc., and the nth member $=a+(n—1)d$. If we put the magnitude of this nth member $=l$ we have :

$$l=a+(n—1)d \tag{1}$$

EXAMPLE 1. What is the 21st member of the series 5, 8, 11... Here $a=5$; $d=3$; $n=21$, and :

$$l=5+(21—1)3$$
$$=5+20.3=65$$

EXAMPLE 2.   What is the 100th member of the arithmetical series 10, 8⅓, 6⅔... etc.?

Here $a=10$; $d=-\tfrac{5}{3}$; $n=100$

$$l=10+(100-1)(-\tfrac{5}{3})$$
$$l=10-165=-155$$

244.

PROBLEM.   Let it be required to find a general formula by which, when the first member $=a$, the last member $=l$, and the number of members $=n$, are given, the sum $=s$ can be determined.

Solution.   If we remember that the series is formed by the constant addition of the difference to each successive member, we perceive that each member is greater, by the difference, than the next preceding member, hence that if we add together any two members which are equally distant from the two ends of the series, the sum will be always the same.   That is, the sum of the first and last members will be exactly the same as the sum of the second and the next to the last, or the third and the third from the last, &c., &c.   If, for example, we take the series 1, 3, 5, 7, 9, 11, 13, 15, 17, 19 and write it twice, the second time in the reverse order, and add the two together, we shall have everywhere the same sum, thus :

1	3	5	7	9	11	13	15	17	19
19	17	15	13	11	9	7	5	3	1
20	20	20	20	20	20	20	20	20	20

If we call the first member a, the last l, and the difference d, the second member is $a+d$, the next to the last $l-d$, the third member is $a+2d$, the third from the last $c-2d$, whence :

$$a+l=a+d+l-d=a+2d+l-2d, \text{ etc.}$$

If, therefore, we add the first and last members of the series together, and multiply the sum by the number of members, we shall have *twice* the sum of the whole series, and this divided by 2, must therefore be the sum of the series.   The desired formula is therefore :

$$s=\left(a+l\right)\frac{n}{2} \tag{2}$$

EXAMPLE 1. What is the sum total of the first 1000 numbers of the so-called "natural" series of numbers 1, 2, 3, ... 1000?

Here $a=1$, $l=1000$, and $n=1000$, whence:

$$s=(1+1000)\tfrac{1000}{2}=500500$$

EXAMPLE 2. What is the sum of the following arithmetical series of ten numbers?

$$6,\ 4,\ 2,\ 0,-2,-4,-6,-8,-10,-12$$

Here $a=6$, $l=-12$, $n=10$

$$s=(6-12)\tfrac{10}{2}=-30$$

245.

In any arithmetical series, we have for consideration, five quantities, a, d, n, l, and s, and the relations of these are given in the two equations :

$$l=a+(n-1)d \qquad (1)$$

$$s=(a+l)\,\frac{n}{2} \qquad (2)$$

Now since these quantities must have the same values in both equations for one and the same series, it follows that when for any series we have three of the five quantities given, we can determine the other two. This follows because we have two equations and only two unknown quantities. Likewise, if we have four of the five quantities given, we can determine the remaining one from the two equations, by eliminating the quantity which is not required, and the other one can then readily be determined.

246.

EXAMPLE 1. It is required to interpolate eight numbers between 4 and 10, in such a manner that the whole shall form an arithmetical series.

Solution. We here have given $a=4$, $l=10$, $n=10$ and d is

if these quantities are contained in equation (1) :

$$l=a+(n-1)d$$

and solving this for d, we get:

$$d = \frac{l-a}{n-1}$$

whence, $d = \dfrac{10-4}{10-1} = \tfrac{2}{3}$

and the reqired series is: 4, $4\tfrac{2}{3}$, $5\tfrac{1}{3}$, 6, $6\tfrac{2}{3}$, $7\tfrac{1}{3}$, 8, $8\tfrac{2}{3}$, $9\tfrac{1}{3}$, 10.

247.

Example 2. A roof has on it 21 rows of tiles; in each row there is one more than in the preceding row, and there are 588 tiles in all. How many are there in the first row?

Solution. Here the given quantities are: $n=21$, $d=1$, $s=588$, and a is required. These quantities are not all in either of the equations, but are in both of them together. The quantity l. being the one which does not enter into the problem, is to be eliminated.

We have the original equations:

$$l = a + (n-1)d \qquad (1)$$

$$s = (a+l)\frac{n}{2} \qquad (2)$$

and can eliminate l, by substituting its value from (1) in equation (2), which gives:

$$s = [a + a + (n-1)d]\frac{n}{2}$$

whence, $\dfrac{2s}{n} = 2a + (n-1)d$

$$2a = \frac{2s}{n} - (n-1)d$$

$$a = \frac{s}{n} - \frac{(n-1)d}{2}$$

whence, $a = \dfrac{588}{21} - \dfrac{(21-1)}{2} \cdot 1 = 118$

248.

Example 3. In a given arithmetical series, we have the first member $a=16$; the difference $d=33$; and the sum $s=1600$. How many members does it contain?

Solution. The three given quantities a, d, and s, and the required quantity n, are contained in both formulas:

$$l = a + (n-1)d \qquad (1)$$

$$s = (a+l)\frac{n}{2} \qquad (2)$$

and l is not required. Eliminating l, by substituting its value from (1) in (2), we have :

$$s=[a+a+(n-1)d]\frac{n}{2}$$

This equation must now be solved for n. We have ($\S220$):

$$2an+n(n-1)d=2s$$
$$n^2d+2an-dn=2s$$
$$n^2+\frac{(2a-d)n}{d}=\frac{2s}{d}$$
$$n=\frac{d-2a}{2d}\pm\sqrt{\left(\frac{d-2a}{2d}\right)^2+\frac{2s}{d}}$$

Whence :

$$n=\frac{32-2\cdot16}{2\cdot32}\pm\sqrt{\left(\frac{32-2\cdot16}{2\cdot32}\right)^2+\frac{2\cdot1600}{32}}$$
$$n=\sqrt{\frac{2\cdot1600}{32}}=10$$

249.

EXAMPLE 4. In an arithmetical progression the first number $=a$; the difference d ; and the sum equals s. Required a general formula for the last number l.

Solution. The required quantity l, appears in both equations :

$$l=a+(n-1)d \tag{1}$$
$$s=(a+l)\frac{n}{2} \tag{2}$$

Eliminating the quantity n, which is not required, by taking its value from (1) and substituting it in (2) we have :

$$n=\frac{l-a}{d}+1$$

whence : $$s=(a+l)\cdot\frac{l-a+d}{2d}$$

and : $$l=-\tfrac{1}{2}d\pm\sqrt{2ds+(a-\tfrac{1}{2}d)^2}$$

II. Geometrical Series.
250.

A series in which we find that the *quotient* obtained by dividing any member into its next following member, is a *constant* quantity, is called a Geometrical Series. Such a series is called an increasing or diminishing series, according as the successive members become uniformly greater or smaller. The constant quotient is called the *exponent* of the series. Examples of geometrical series are as follows :

$$3, \quad 6, \quad 12, \quad 24, \quad 48, \quad 96 \ldots$$
$$9, \quad 3, \quad 1, \quad \tfrac{1}{3}, \quad \tfrac{1}{9}, \quad \tfrac{1}{27} \ldots$$
$$\tfrac{1}{2}, \quad \tfrac{1}{4}, \quad \tfrac{1}{8}, \quad \tfrac{1}{16}, \quad \tfrac{1}{32}, \quad \tfrac{1}{64} \ldots$$

In the first part of the above series, the exponent is $=2$, in the second it is $\tfrac{1}{3}$, this being a diminishing series, and in the third the exponent $=\tfrac{1}{2}$.

251.

When the first member and the exponent of a geometrical series are given, the series can easily be constructed. The second member is obtained by multiplying the first by the exponent, the third is obtained by multiplying the second member by the exponent, or what is the same thing, by multiplying the first member by the second power of the exponent, etc., so that the 99th member multiplied by the exponent gives the 100th member, or the first member multiplied by the 99th power of the exponent also gives the 100th member.

If for example the first member $=2$, and the exponent $=3$, we have the series :

$$\begin{array}{cccccc} (1) & (2) & (3) & (4) & (5) & (\text{10th member}) \\ 2 ; & 2 \cdot 3 ; & 2 \cdot 3^2 ; & 3 \cdot 3^3 ; & 2 \cdot 3^4 \ldots \ldots 2 \cdot 3^9 \ldots \ldots \\ \text{or} \quad 2 ; & 6 ; & 18 ; & 54 ; & 162 ; \ldots \ldots 39366 \ldots \ldots \end{array}$$

If the first member $=64$, and the exponent $=\tfrac{1}{2}$, we have for the series :

$$\begin{array}{ccccc} (1) & (2) & (3) & (4) & (10) \\ 64 ; & 64 \cdot \tfrac{1}{2} ; & 64 \cdot (\tfrac{1}{2})^2 ; & 64(\tfrac{1}{2})^3 \ldots 64 \cdot (\tfrac{1}{2})^9 \ldots \\ \text{or} \quad 64 ; & 32 ; & 16 ; & 8 ; \ldots \ldots \tfrac{1}{8} \ldots \ldots \end{array}$$

252.

In geometrical series, as in arithmetical series, there are five quantities with their symbols, to be considered, namely : the first member a, the exponent e, the number of members n, the last member l, and the sum s.

Each of these five quantities is a determinate function of any three of the others, and if three of the five are given, the other two can be determined without requiring the series to be constructed.

The most important questions are the determination of the magnitude of any particular member, and the determination of the sum of all the members.

253.

PROBLEM. Required to find the algebraic expression from which the value of any particular member l can be found, when we have given its number in the series $= n$, also the first member $= a$, and the exponent $= e$.

Solution. According to § 251, the series is :

$$
\begin{array}{cccccc}
(1) & (2) & (3) & (4) & (5) \ldots .(n-1) \\
a\,; & ae\,; & ae^2\,; & ae^3\,; & ae^4\,; \ldots .ae^{n-1}
\end{array}
$$

whence : $l = a \cdot e^{n-1}$

or in words : *In order to find the value of the nth member of a geometrical series, raise the exponent to the n—1st power, and multiply it by the first member.*

If n is very large, the labor will be greatly reduced by the use of logarithms, as will be explained in Book XXI. Until then, we shall only give such examples as are easily solved without the aid of logarithms.

EXAMPLE. The first member of a progression is $\frac{1}{64}$, the exponent is 2 ; how large is the 9th member ?

Solution. Here $a = \frac{1}{64}$, $e = 2$, $n = 9$, and l, is required.

$$l = a \cdot e^{n-1}$$
$$l = \tfrac{1}{64} \cdot 2^{9-1} = \tfrac{1}{64} \cdot 2^8 = 4$$

254.

PROBLEM. Required to find the formula from which the sum s, can be found (the "summation formula"), when we have given the first member a, the exponent e, and the last member l.

Solution. The solution of this problem requires the exercise of a little judgment. If we call the sum of a geometrical progression, such as $a + ae + ae^2 + ae^3 + \ldots l$ equal to s, we have :

$$s = a + ae + ae^2 + ae^3 + \ldots + \frac{l}{e^2} + \frac{l}{e} + l \qquad (1)$$

Multiplying both sides of this equation by the exponent e, we have :

$$es = ae + ae^2 + ae^3 + ae^4 + \ldots + \frac{l}{e} + l + le \qquad (2)$$

Then subtracting equation (1) from equation (2) we get :

$$es - s = le - a$$

$$(e-1)s = le - a$$

$$s = \frac{le - a}{e - 1} \qquad (3)$$

Expressed in words this is : *To find the sum of a geometrical series, multiply the last member by the exponent and subtract the first member, then divide this by the exponent less one.*

This expression for "s" is important in subsequent work and should be carefully remembered.

EXAMPLE. What is the sum of the following progression :

$1, \frac{1}{2}, \frac{1}{4}, \frac{1}{8}, \frac{1}{16}, \frac{1}{32}, \frac{1}{64}, \frac{1}{128}, \frac{1}{256}, \frac{1}{512}, \frac{1}{1024}$ (see §329)

Solution. $a = 1$, $e = \frac{1}{2}$, $l = \frac{1}{1024}$, s is required.

$$s = \frac{le - a}{e - 1}$$

$$s = \frac{\frac{1}{1024} \cdot \frac{1}{2} - 1}{\frac{1}{2} - 1} = \frac{-\frac{2047}{2048}}{-\frac{1}{2}}$$

$$s = 1\frac{1023}{1024}$$

255.

From the following fundamental formulas of geometric series:

$$l = a \cdot e^{n-1} \qquad \text{I.}$$

$$s = \frac{le - a}{e - 1} \qquad \text{II.} \quad \Big\}$$

we can determine formulas for the values of any of the quantities a, e, n, l, s, when any three are given, in a similar manner to that shown in §245. Some of the resulting problems require the use of logarithms for their solution, others lead into mixed equations of the higher degrees, the solution of which is taught in Higher Algebra or Analysis.

256.

Geometrical series expressed in letters can be readily summed up by means of formula II.

Thus suppose we have the series :

$$b + bz + bz^2 + bz^3 + bz^4 + \ldots + bz^{n-1}$$

in which b is the first term, bz^{n-1} the last term, and z the exponent.

Substituting these for a, l, and b, in the formula

$$s = \frac{le - a}{e - 1}$$

we have :

$$b + bz + bz^2 + bz^3 + bz^4 + \ldots \; bz^{n-1} = \frac{b(z^n - 1)}{z - 1}$$

or of the series :

$$1 + e + e^2 + e^3 + e^4 + \ldots \; e^{n-1} = \frac{e^n - 1}{e - 1} = \frac{1 - e^n}{1 - e} \; ;$$

$$1 - x + x^2 + x^3 + x^4 + \ldots \; + x^{2n} = \frac{-x^{2n+1} - 1}{-x - 1} = \frac{1 + x^{2n+1}}{1 + x} \; ;$$

$$y + x + \frac{x^2}{y} + \frac{x^3}{y^2} + \frac{x^4}{y^3} + \ldots \; + \frac{x^n}{y^{n-1}} = \frac{x^{n+1} - y^{n+1}}{(x - y)y^{n-1}} \, .$$

BOOK XIX.

Logarithms.

257.

Before logarithms were discovered and complete tables of
logarithms computed, practical arithmetic may be said to have
come to a standstill. To-day, calculations, which in the time
of Kepler required days and weeks for their completion, or were,
to the great loss of mankind, abandoned because of the im-
mense labor involved in their computation ; to-day, such calculations may be made in a few minutes by the use of logarithms,
even by the veriest beginner in mathematics. It has well been
said that logarithms are to arithmetic what the steam engine is
to mechanics.

In order to obtain a clear notion of the principle on which
logarithms are based, let us take any number, such as 2, for example, and raise it to a series of powers beginning from o, thus :

$$1 = 2^0, \quad 2 = 2^1, \quad 4 = 2^2, \quad 8 = 2^3, \quad 16 = 2^4, \text{ etc.}$$

Then in order better to examine the relations which exist
between these powers and their exponents, let us write them in
vertical columns with a dividing line between. The base number
2 and the sign of equality we may omit as being the same for all
of the powers here given. Thus instead of $1 = 2^0$, $128 = 2^7$, we
shall write more briefly, $1 \mid 0$, $128 \mid 7$, etc.

Powers	Exponents	Powers	Exponents	Powers	Exponents
1	0	1024	10	1048576	20
2	1	2048	11	2097152	21
4	2	4096	12	4194304	22
8	3	8192	13	8388608	23
16	4	16384	14	16777216	24
32	5	32768	15		
64	6	65536	16		
128	7	131072	17		
256	8	262144	18		
512	9	524288	19		

258.

Now remembering the four fundamental rules for powers, (§ 204 to 208),

$$a^m . a^n = a^{m+n} \qquad (1)$$

$$\frac{a^m}{a^n} = a^{m-n} \qquad (2)$$

$$(a^n)^m = a^{mn} \qquad (3)$$

$$\sqrt[n]{a^m} = a^{\frac{m}{n}} \qquad (4)$$

we see how such a table of powers and exponents can be made of use.

With the exception of addition and subtraction, any operations desired upon any numbers in the first column can be greatly abridged if we take instead the corresponding exponents, and thus convert multiplication into addition, division into subtraction, the raising of powers into multiplication and the extraction of roots into simple division. All these can be accomplished by such a simple, but invaluable table of powers and exponents.

EXAMPLE 1. If two or more numbers, such as 128 and 512 are to be multiplied together, we look for these factors in the first column, and take the corresponding exponents and *add* them together ; then find in the column of exponents one which is equal to their sum and opposite it will be the desired product. Thus we have it in our little table :

$$\text{Exponent of } 128 = 7$$
$$\text{" } \text{" } 512 = 9$$
$$\overline{\hspace{1.5em} 16}$$

Opposite the exponent 16 we find the power 65536, which is the product of the two factors.

The reason for this is readily seen. All the numbers which are in the first columns are powers of the same base number. Thus in the above example :

$$128 = 2^7$$
$$512 = 2^9$$

But $128 \cdot 512 = 2^7 \cdot 2^9 = 2^{16}$ (see § 204), hence that power of 2 which has the exponent 16 must be the product of the given numbers, or

$$128 \cdot 512 = 65536.$$

EXAMPLE 2. Suppose it is required to divide any of the numbers in the first column by any other of these, we merely *subtract* the exponent of the divisor from the exponent of the dividend, and the power which corresponds to the resulting difference, will be the quotient.

Thus we have :

$$\frac{2097152}{256} = 8192$$

Because :

Exponent of $2097152 = 21$
"   " $256 = 8$

Difference   13
$8192 =$ power whose exponent is $= 13$.

This is the same as :

$$\frac{2097152}{256} = \frac{2^{21}}{2^8} = 2^{21-8} = 2^{13} = 8192 \qquad \text{(see § 205)}$$

EXAMPLE 3. If it is required to raise any quantity in the first column to any given power, we simply *multiply* the exponent in the table by that of the required power, and opposite the product we find the required power. Thus :

$$16^5 = 1048576$$

We have: Exponent of $16 = 4$
Multiplied by   5
———
20

And the power whose exponent in the table is 20, is 1048576.

For $16 = 2^4$
whence $16^5 = (2^4)^5 = 2^{20} = 1048576$ (see § 207)

EXAMPLE 4. If it is required to extract the root of any number in the first column, we simply divide the exponent of the given number by the exponent of the required root, and the power corresponding to the quotient is the required root. Thus:

$$\sqrt[7]{2097152}=8$$

This follows because :

Exponent of $2097152=21$

$$\frac{21}{7}=3$$

Power whose exponent is 3, is 8,

or $2097152=2^{21}$

$$\sqrt[7]{2097152}=\sqrt[7]{2^{21}}=2^{\frac{21}{7}}=2^3 \qquad \text{´(See § 208)}$$

259.

We see the great value even of such an imcomplete table as that given in the preceding paragraph, but its usefulness is very limited because of the great gaps which exist in it. If we desire to make calculations involving numbers between 2 and 4, 4 and 8, 8 and 16, etc., we are unable to use the table, because it does not contain these numbers.

When, however, these numbers are all introduced, and their corresponding exponents calculated, then the system becomes complete and is of the highest degree of utility for all calculations. Such a table is called a Table of Logarithms. The quantities in the first column are called the *Numbers*, while their corresponding exponents in the second column are called the *Logarithms*. We have also to discuss in this connection the base number upon which the exponents are calculated, and this will be referred to as the *Base*.

Notwithstanding the great value of such tables of logarithms, no private individual could take the time and labor to calculate a table for himself. This gigantic work has, however, been very fully and accurately done already, and many excellent editions of logarithms published.

The best tables are those calculated to seven decimal places, and the best of these are the magnificent foreign editions of Schrön and of Bruhns. The American edition of Stanley is also good. For many purposes six decimals aie sufficient, and good six figure tables are those of Loomis, published by Harper & Bros., and those published by Prof. Geo. W. Jones at Ithaca, N. Y. Very often five decimals are quite enough for ordinary calculations and the five figure logarithms contained in Trautwines Tables, will be found very useful.

Logarithms were invented in the early part of the 17th century by Baron Napier, of Merchiston, of Scotland. The first full table of logarithms was calculated shortly afterward by Henry Briggs, also of Scotland. the work requiring an entire year and the help of eight assistants. This table was to 14 decimal places, and for all numbers up to 10,000. It was afterwards extended to all numbers up to 100,000, by Adrian Vlacq, a Hollander. Vlacq's tables were to 10 decimal places, and on these original tables, with many errors corrected, the modern extremely accurate tables are all based.

260.

It is evidently immaterial, so far as the principle is concerned, what number is taken as the *base* for a system of logarithms. When the system is once made, the base is no longer used. In the little table in § 257 the base is 2. Any other number may be taken, however, and the resulting system will be equally useful. Since the base may be chosen at will, all the tables above referred to, derived from the original tables of Briggs, have been calculated on the same base as our number system, i. e., 10.

Logarithms calculated to the base 10, are called *common* logarithms, or sometimes, Briggs' logarithms. There is also another, the so-called Natural System (also called Hyperbolic Logarithms), which, although not so convenient in practice is of the greatest importance in higher mathematics, and in steam engineering.

In the following pages it is assumed that the student has provided himself with a good set of logarithmic tables, and as the following problems are all worked with seven-figure logarithms, he had better procure a good set, such as Schrön or Bruhns, although the cheaper sets will answer if these are not at hand. For the student of engineering or physical science it is very desirable that he should provide himself with a good set of logarithms and acquire every possible facility in their use, and the familiarity which one obtains with the pages of a special, personal copy is such as to make it desirable that he should get at the beginning the set he intends to keep as his companion in all his subsequent scientific work.

In nearly all sets of logarithmic tables instructions and explanations for their use are also given, but as these are not always expressed as clearly as might be desired, we will here proceed to explain them in a full and practical manner. No explanations, however, can take the place of frequent practical use, and the greatest facility in the use of such tables must be obtained by actual practice.

261.

In examining a set of seven-figure common logarithms we find among the various values the following, which will serve as an explanation :

NUMBER	LOGARITHM		NUMBER	LOGARITHM
1	0.0000000		10	1.0000000
2	0.3010300			
3	0.4771213		99	1.9956352
4	0.6020600		100	2.0000000
5	0.6989700			
6	0.7781513		999	2.9995655
			1000	3.0000000
			9999	3.9999566

According to § 257 these numbers simply mean that $1 = 10^0$;

$$2 = 10^{0.301030} = 10^{\frac{301030}{1000000}} = \sqrt[1000000]{10^{301030}}$$

$3 = 10^{0.4771213}$, etc.; $10 = 10^1$; $100 = 10^2$, etc.

The Briggs logarithms are therefore nothing more than the *exponents* of those powers to which the number 10 must be raised in order to produce the corresponding *numbers* opposite them in the table. The sign of equality, and the base 10, are omitted from the tables to save room, and are always understood. We see that 0 is the logarithm of 1, since the only power of 10, which is equal to 1, is the 0th power ; the logarithm of 2 is 0.3010300, that being the power to which 10 must be raised, to equal 2 ; the logarithm of 10 is 1, since the 1st power of 10=10. Therefore we may write :

log. 1=0.0000000 ; log. 10=1.0000000 ; log. 2=0.3010300 ; log. 99=1.9956352, etc.

262.

In the higher Analysis, short methods are derived by means of which a skilled computer can calculate logarithms very rapidly.

Since the student is not yet prepared to study those methods, we shall here explain the more tedious arithmetical method by which they may be calculated. This method is the one by which Briggs calculated his tables, the modern system of higher mathematics not having been invented in his time.

In order to calculate the logarithm of any number, such as 5, for example, we proceed as follows :

The number 5, considered as a power of 10, lies somewhere between 10^0, and 10^1, since $10^0 < 5$, and $10^1 > 5$.

Hence we think that it *may* be equal to 10 raised to the $\frac{0+1}{2} = \frac{1}{2}$ power. We find by trial that $10^{\frac{1}{2}} = \sqrt{10} = 3.1622776601$. This is too small, for $10^{\frac{1}{2}} < 5$; so the logarithm must lie between the narrower limits of $\frac{1}{2}$ and 1, since $10^{\frac{1}{2}} < 5$, and $10^1 > 5$. In this manner we can reduce the limits narrower and narrower, by taking repeatedly the half sum of the greater and lesser exponents, and yet never have to extract a higher root than the square root. Thus, suppose now we try the $\frac{\frac{1}{2}+1}{2} = \frac{3}{4}$ power of 10. We have (§ 210) :

$$10^{\frac{3}{4}} = \sqrt{10^{\frac{3}{2}}} = \sqrt{(10^1 \cdot 10^{\frac{1}{2}})} = \sqrt{10 \cdot (3.1622776601)} = 5.6234132$$

hence, $10^{\frac{1}{2}} < 5$, and $10^{\frac{3}{4}} > 5$

Again :

$$10^{\frac{\frac{1}{2}+\frac{3}{4}}{2}} = 10^{\frac{5}{8}} = \sqrt{10^{\frac{5}{4}}} = \sqrt{(10^{\frac{3}{4}} \cdot 10^{\frac{1}{2}})} =$$

$$= \sqrt{(5.623\ldots)(3.1622\ldots)} = 4.21695034$$

hence $10^{\frac{5}{8}} < 5$, and $10^{\frac{3}{4}} > 5$

Again :

$$10^{\frac{\frac{5}{8}+\frac{3}{4}}{2}}=10^{\frac{11}{16}}\sqrt{(10^{\frac{5}{8}}\cdot10^{\frac{3}{4}})}=\sqrt{(4.2169..)}\,(5.623..)=4.869675252$$

hence, $10^{\frac{11}{16}}<5$, and $10^{\frac{3}{4}}>5$

Again :

$$10^{\frac{\frac{11}{16}+\frac{3}{4}}{2}}=\sqrt{(10^{\frac{11}{16}}\cdot10^{\frac{3}{4}})}=5.232991,\text{ etc.}$$

We thus see that the approximation is growing constantly nearer and nearer and after repeating the operation 22 times we find that : $10^{\frac{29311883}{42581047}}=5.00000086$, or $=5$ correct to the sixth decimal, and hence log. $5=\frac{29311883}{42581047}=0.6989700$, it being much more convenient to have the logarithm expressed as a decimal.

Logarithms cannot be calculated with absolute exactness, because they belong to the so-called irrational quantities, of which the decimals are endless.

263.

If the original tables of Briggs and Vlacq had not already been calculated by the above tedious process, the work might have been greatly reduced by means of modern methods. It is only necessary in this way first to compute the logarithms of the prime numbers, since all the others can be derived from them by simple addition and multiplication. This follows, because all composite numbers can be resolved into their prime factors. If we know the logarithms of the various factors of a number, we have simply to add the logarithms of the factors together to obtain the logarithm of the product. Thus if we add together the logarithms of 2 and 3, we obtain the logarithm of 6. The reason for this is readily seen if we call log. $2=a$, and log. $3=b$, whence $2=10^a$ and $3=10^b$, whence $10^a\cdot10^b=10^{a+b}$.

By comparison with the table of page 257 we see how the addition of the logarithms of 2 and of 3 gives the logarithm of 6, and we must observe carefully the following points.

In the calculation of the original tables, more than seven decimal places were computed. Of these the first seven were retained, in the preparation of seven figure tables, and the last decimal was increased by 1 whenever the 8th decimal was greater than 5. For this reason exact dependence cannot be placed upon the accuracy of the last decimal. This variation from exactness can, however, be neglected to a great extent in actual practice, since the influence of the last decimal of the logarithm has such a slight influence upon the result as to be of little importance. If we have many logarithms to add together, or have to multiply a logarithm by a large number, the result will come out too large if the last figure is too large. But if, as is often the case, we have again to divide this resulting logarithm by another number, the error will again be reduced. In some tables, such as Schrön's, there is a mark placed under the last decimal whenever it has been so increased, and by taking this into account a correction can be made when extreme accuracy is necessary.

264.

In the Briggs system of logarithms, of which the base number is 10, we have the logarithm of 10=1, and the logarithm of 1=0, hence for all numbers between 1 and 10, the logarithms must be greater than 0 and less than 1, i. e., must be proper fractions. Also, since the logarithm of 100 is 2, the logarithms of all numbers between 10 and 100 must be >1 and <2, and must be mixed numbers, or consist of a whole number and a decimal. That portion of a logarithm which consists of a whole number, is called the *characteristic*, while the decimal part is called the *mantissa*, and as these names will frequently be used hereafter, care should be taken to remember them. In the common system, the logarithms of all numbers except those which are even powers of 10, are mixed numbers, and *the characteristic of any logarithm is always 1 less than there are figures in the number.*

For all numbers of one figure, the characteristic is 0, for all numbers of two figures the characteristic is 1, for all 3 figure numbers it is 2, etc., etc. Since the characteristic of the logarithm of any number can thus always be obtained at once by inspection of the number, it is unnecessary to have the characteristics in the tables ; and they are therefore omitted in order to save room. In looking in a table, therefore, such as Schrön's or Bruhn's, we find only the decimal part of the logarithm, and must prefix the characteristic ourselves by the above rule. For example :

log. 4571=3.6600112
log. 4577=3.6605809, etc.

265.

Since by adding together the logarithms of several factors we obtain the logarithm of their product, and since the logarithms of all the simple units (10, 100, 1000, etc.), have characteristics which are whole numbers, and mantissas=0, i. e., log. 1=0, log. 10=1, log. 100=2, etc., it is clear that when any number is multiplied by 10, 100, 1000, etc., the mantissa of its logarithm remains unchanged, and only the characteristic is altered.

If, therefore, we know the logarithm of 2, we know also the logarithm of 20=2·10, or of 200=2·100, etc.

From the logarithm of 47, we have merely by changing the characteristic, the logarithms of 470, 4700, 47000, etc.

Thus we find by reference to the tables :

log.	2=0.3010300 ;	also log.	47=1.6720979
"	20=1.3010300 ;	"	470=2.6720979
"	200=2.3010300 ;	"	4700=3.6720979
"	2000=3.3010300 ;	"	47000=4.6720979
"	200000=5.3010300 ;	"	47000000=7.6720979

266.

The logarithm of a quotient, is obtained by subtracting the logarithm of the divisor from the logarithm of the dividend. For this purpose, also, we consider all fractions as expressions of division. It therefore follows that when any number is to be divided by 10, 100, 1000, etc., we simply reduce the characteristic of its logarithm by as many units as there are zeros in the divisor and leave the mantissa unchanged.

Thus, for example :

$$\text{log. } 4571=3.6600112$$
$$\text{whence, } \text{log. } \tfrac{4571}{10}=2.6600112$$
$$\text{log. } \tfrac{4571}{100}=1.6600112$$

This follows because :

$$\tfrac{4571}{100}=\frac{10.^{3.6600112}}{10^2}=10^{1.6600112}.$$

267.

From the foregoing paragraphs we can deduce at once the rule by which to find the logarithm of any whole number and decimal combined. Take first the characteristic belonging to the *whole* part of the number *only*. Then take from the tables the mantissa which corresponds to the number *including* the decimal *portion* as if there were no decimal point present. Thus :

$$\tfrac{4571}{10}=457.1 \; ; \; \tfrac{4571}{100}=45.71, \text{ etc.}$$

Therefore : log. $457.1=2.6600112$

" $45.71=1.6600112$

" $4.571=0.6600112$

268.

When we have to find the logarithm of a whole number and common fraction, such for example as $36\tfrac{3}{4}$, we may proceed in two ways. We may convert the fraction into a decimal and proceed as in the preceding paragraph, or we may convert it into an improper fraction, and then subtract the logarithm of the denominator from the logarithm of the numerator.

Thus, since $36\tfrac{3}{4}=36.75=\tfrac{147}{4}$, we have :

$$\text{log. } 36\tfrac{3}{4}=\text{log. } 36.75=1.5652573$$

$$\text{or } \text{log. } 36\tfrac{3}{4}= \left\{ \begin{array}{l} \text{log. } 147=2.1673173 \\ \quad\text{"} \quad\; 4=0.6020600 \end{array} \right.$$

$$\text{log. } \tfrac{147}{4}=1.5652573$$

269.

To take out the logarithms of proper fractions or of decimals from the table and determine their correct characteristics, we we must observe the following points : Since $10^0=1$, the exponent for any quantity less than 1 must be smaller than 0, and must be a proper fraction. Thus, for example :

$$10^{-1}=\tfrac{1}{10}, \; 10^{-2}=\tfrac{1}{100}. \qquad \text{(See § 205).}$$

The logarithms of all proper fractions must therefore be negative, and the smaller the value of the proper fraction the greater the absolute number of the corresponding negative logarithm.*

---

*When a fraction becomes infinitely small its negative logarithm must become infinitely large. A magnitude which becomes infinitely large, and can no longer be expressed by figures, is indicated by the sign ∞; and a quantity which has become infinitely small, so as to be indistinguishable from o, is represented by $\tfrac{1}{\infty}$. These expressions also give rise to:

$$10^{-\infty}=\tfrac{1}{10^\infty}=0=\tfrac{1}{\infty}$$

Therefore we have : log. $o=-\infty$ (Read "infinite negative",.

In order to write the negative logarithm of a true decimal (i. e., a decimal without any whole number attached) acording to § 266, we can unite the decimal with its proper denominator, and then subtract the logarithm of the denominator from the logarithm of the numerator.

Thus for example :

$$0.0564 = \tfrac{564}{10000}$$

$$\log. 564 = \quad 2.7512791$$
$$\text{and log. } 10000 = \quad 4.0000000$$
$$\text{hence log. } 0.0564 = -1.2487209$$

$$\text{for } \tfrac{564}{10000} = \frac{10^{2.7512791}}{10^4} = 10^{2.7512791-4} = 10^{-1.2487209}$$

270.

When a negative logarithm occurs, not as the final result of a calculation, but is to be added to or subtracted from other logarithms, or the number corresponding to it is to be determined, it is then found much more convenient to separate the negative logarithm into two parts, of which one part (the mantissa) shall consist of a *positive*, proper decimal fraction, and the other part a *negative* whole number, connected to the positive part as a negative characteristic.

A negative logarithm is easily brought into this form, by appending such a positive number, in connection with + and — signs, as shall convert the logarithm into one with a negative characteristic, and positive mantissa. This is done in the following manner :

Suppose, for example, that we have :

$$\log. 0.0564 = -1.2487209$$

we may add 2 and subtract 2, by which operation the actual value of the logarithm will not be altered, but only its form changed. Thus :

$$\log. 0.0564 = 2 - 1.2487209 - 2$$

Then collecting the first two members of the right-hand side together we have :

$$\log. 0.0564 = 0.7512791 - 2$$

giving us the required form, in which we have the positive mantissa with o for a characteristic, and also a negative whole number

(—2) for the negative characteristic. Instead of writing the negative characteristic after the mantissa, it is often written before it, in the usual place, but with the minus sign *over* the characteristic, in order to show that the whole number only is negative, and that the minus sign does not apply to the decimal portion, thus :

$$\log. \; 0.0564 = \overline{2}.7512791$$

271.

The denominator of a decimal fraction is equal to 1, with as many zeros appended as there are places in the decimal, thus :

$$0.564 = \tfrac{564}{1000}; \quad 0.0564 = \tfrac{564}{10000}, \text{ etc.}$$

We therefore see that the characteristic of the logarithm of the denominator will be greater than that of the numerator by as many units as there are zeros between the decimal point and the first significant figure of the decimal. Thus in the fraction $\tfrac{564}{1000}$ the characteristic of the numerator will be 2 and that of the denominator will be 3 ; while for $0.0564 = \tfrac{564}{10000}$, the characteristic of the denominator will be 4, while that of the numerator will still remain $= 2$.

From these considerations we obtain a simple rule to find the logarithm of a proper decimal fraction from the regular tables. We first find the logarithm of the number of which the decimal is composed, just as if the decimal point did not exist, and write it with a characteristic $=0$, and then append a negative characteristic of as many units as there are zeros in the decimal plus 1. Thus :

$$\log. \quad 0.564 = 0.7512791 - 1, \text{ or } \overline{1}.7512791$$
$$\text{``} \quad 0.0564 = 0.7512791 - 2, \text{ or } \overline{2}.7512791$$
$$\text{``} \quad 0.00564 = 0.7512791 - 3, \text{ or } \overline{3}.7512791$$

This operation simply amounts to considering the decimal as a vulgar fraction with its corresponding denominator of 100, 1000, etc., and then taking the logarithm of the numerator and subtracting that of the denominator, except that the subtraction is *indicated* instead of being performed.

272.

In order to find the logarithm of a proper vulgar fraction, the fraction may be converted into a decimal and the logarithm found at once by the above rule. Thus we have:

$$\log. \tfrac{7}{16} = \log. \ 0.4375 = 0.6409781 - 1$$
$$\text{or,} \quad = \overline{1}.6409781$$

Frequently it is more convenient to subtract the logarithm of the denominator from the logarithm of the numerator, as the following example shows:

$$\log. \ 7 = 0.8450980$$
$$\text{`` } \ 16 = 1.2041200$$
$$\log. \tfrac{7}{16} = \overline{1}.6409780$$

Here we have a positive mantissa, being the difference between the two mantissas, while the characteristic is $0-1 = -1$, as before.

The practice of placing the characteristic in the usual place, with the negative sign over it, is much the more convenient in practice, and will be used altogether hereafter.

Again we have:

$$\log. \tfrac{3}{7} = \begin{cases} \log. \ 3 = 0.4771213 \\ \log. \ 7 = 0.8450980 \end{cases}$$
$$\overline{1}.6320233$$

Here the lower decimal is greater than the upper, and hence 1 was borrowed from the unit's place, which gives —1 in the result.

Again:

$$\log. \tfrac{11}{4771} = \begin{cases} \log. \quad 11 = 1.0413927 \\ \log. \ 4771 = 3.6786094 \end{cases}$$
$$\overline{3}.3627833$$

Here the lower decimal is the greater, and 1 is borrowed from the unit's place of the subtrahend, leaving 0, and then $0-3 = -3$ for the negative characteristic.

Whenever circumstances may require it, we can modify any logarithm by adding, *at the same time*, equal positive and negative characteristics. Thus:

$$\log. \tfrac{3}{7} = \quad \overline{1}.6320233$$
$$-6+6$$
$$= -6 + 5.6320233$$

or   log. $\frac{3}{4}$=   1.6320233

                        —4+4

                —4+3.6320233 etc.

The use of this will be seen hereafter.

273.

Bearing in mind the rules given in the preceding paragraphs we will now proceed to explain more fully the method of taking out from the tables the logarithms of any required numbers.

If we open any table of seven figure logarithms, such as Schrön's, we see at the left of the page the column headed Num. (Numbers). Then following toward the right 10 columns, headed successively 0, 1, 2, 3, 4, 5, 6, 7, 8, 9, and at the right a column headed P. P. (Proportional Parts).

In some tables the column of numbers is simply headed N, and the column of proportional parts is headed, "Differences."

(1) Turning now, for example, to the number 1267 in the column "Num." we find opposite 1267 in the column 0, the logarithm .1027766, or with the characteristic, 3.1027766. We do not find the first three figures of the decimal repeated for every number, but only printed in where they change, hence when there is a space for the first three figures those which are next above are to be taken. Likewise the succeeding columns contain only four figures, and these are in each case to be preceded by the corresponding three figures printed in the 0 column. This saves the constant repetition of the first three figures, and economizes space very much.

If a number contains only four figures its logarithm is found at once opposite to the number in the 0 column, but if the number consists of five figures the fifth figure must be found in the column headings at the top of the page, and in the corresponding column and on the same horizontal line as the first four figures will be found the proper logarithm. Thus we have :

log.   1267=3.102 7766   log. 12675=4.102 9480
log. 12670=4.102 7766   log. 12676=4.102 9822
log. 12671=4.102 8109   log. 12677=4.103*0165
log. 12672=4.102 8452   log. 12678=4.103*0507
log. 12673=4.102 8974   log. 12679=4.103*0850
log. 12674=4.102 9137

Examining these we see that the logarithms of 1267 and 12670 have the same mantissa, the characteristic being simply increased by 1. If the fifth figure, however, is greater than 0 we take out the last four decimals from the column headed by this figure, the first three figures of the mantissa remaining unchanged. In some cases, as log. 12677, the third figure also is changed, and this is indicated by an asterisk (*) in front of the fourth figure. When this * appears the first three figures of the mantissa are to be taken from the next line *below* in the zero column. Thus when we look for the log. 12677 we have 102 in the 0 column, and *0165 in column 7, and the asterisk means that instead of 102 we must take 103 for the first three figures. In some tables a dot is used instead of an asterisk, and in others a dash (—), but the meaning is the same in all.

We therefore have the following rule to take out the logarithms of five figure numbers.

First write the proper characteristic, then find the first four figures of the given number in the column "Num," and opposite these in the 0 column find the first three figures of the mantissa. Then find the last four figures of the mantissa in the column headed by the fifth figure of the given number, on the same horizontal line as the first four figures were found. If there is also an asterisk here found, the third figure of the mantissa is to be increased by 1.

If the given five figure number is multiplied or divided by 10, 100, 1000, etc., the mantissa remains unchanged and the characteristic alone is altered. Thus:

$$\log. 22035 = 4.3431131$$
$$\log. 2.2035 = 0.3431131$$
$$\log. 33829 = 4.5292892$$
$$\log. 338.87 = 2.5300331$$
$$\log. 338.29 = 2.5292892$$
$$\log. 78.164 = 1.8930068$$
$$\log. 781640 = 5.8930068$$
$$\log. 0.049097 = 2.6910550$$
$$\log. 1.1011 = 0.0418268$$

274.

If we take out the logarithms of several successive five-figure numbers, and subtract them from each other, we find the differences appear only in the last three figures, and that for some little space these differences are all equal to each other.

Thus for example :

$$\begin{array}{lll}
\log. 23740 = 4.3754807 & \text{Difference} \\
\log. 23741 = 4.4754990 & 183 \\
\log. 23742 = 4.3755173 & 183 \\
\log. 23743 = 4.3755356 & 183 \\
\log. 33744 = 4.3755539 & 183 \\
\log. 23745 = 4.3755722 & 183
\end{array}$$

We thus see that if we increase the number 23740 by 1, we increase the last 3 decimals of its logarithm by 183, and if the number 23740 increases 2, 3, or 4 units, we must add 2, 3, or 4 times 183 to the last figures of its logarithm.

Since this proportion exists for the increase of the number by units, it must also hold good for fractional parts of units, such as $\frac{1}{10}$, $\frac{2}{10}$, $\frac{1}{100}$, $\frac{2}{100}$, etc. If, therefore, the number 23743, for example, is increased by $\frac{1}{10}$, $\frac{2}{10}$, $\frac{1}{100}$, etc., then also will the last decimals of the logarithm be increased by $\frac{1}{10}$, $\frac{2}{10}$, $\frac{1}{100}$, etc., of the difference, namely by $\frac{1}{10} \cdot 183$, or $\frac{2}{10} \cdot 183$, etc.

If therefore we know the difference between the logarithm of 23743 and the next logarithm, we can find the logarithms of $23743\frac{8}{10} = 23743.8$; or of 23743.85 ; or of 23743.859 etc., or of a number 10, 100, or 1000 times as great, such as $237438 = 10 \cdot 23743.8$; or $2374385 = 100 \cdot 23743.85$.

Suppose for example that we have :

$$\log. 23743 = 4.3755356$$

If now we add to the last decimals 0·8 times the difference 183, i. e.,

$$0.8 \cdot 183 = 146.4, \text{ say } 146$$

We have :

$$\begin{array}{r}
4.3755356 \\
146 \\
\hline
\log. 23743.8 = 4.3755502
\end{array}$$

and $\log. 237438 = 5.3755502$

If we add 0.85 times the difference 183, we have:

$$0.85 \cdot 183 = 155.55, \text{ say } 156$$

and

$$\begin{array}{r}
4.3755356 \\
156 \\
\hline
\log. 23743.85 = 4.3755512
\end{array}$$

and $\log. 2374385 = 6.3755512$ etc.

In the best seven figure tables, such as Schrön's, this calculation is rendered unnecessary, for in the column headed P. P. (at the extreme right) are given the products of the differences by the digits from 1 to 9, so that the amount to be added for any required number can be taken at once.

Thus we find for

$$\log. 23743859 = 7.3755513$$

in the following manner :

$$\text{The } \log. 23743 = 4.3755356$$

then for the remaining figures we have :

$$\tfrac{859}{1000}\cdot 183 = (\tfrac{8}{10} + \tfrac{5}{100} + \tfrac{9}{1000})183 =$$
$$= \tfrac{8}{10}\cdot 183 + \tfrac{1}{10}\cdot\tfrac{5}{10}\cdot 183 + \tfrac{1}{100}\cdot\tfrac{9}{10}\cdot 183$$

In the column of Proportional Parts we now find :

$$\tfrac{8}{10}\cdot 183 = 146.4$$
$$\tfrac{5}{10}\cdot 183 = 91.5 \text{ hence } \tfrac{1}{10}\cdot 91.5 = 9.15$$
$$\tfrac{9}{10}\cdot 183 = 164.7 \text{ hence } \tfrac{1}{100}\cdot 164.7 = 1.647$$

and we have :   log. 23743 . . . =4.3755356

$$
\begin{array}{ll}
8 & 146.4 \\
5. & 9.15 \\
9 & 1.645 \\
\end{array}
$$

$$\log. 23743.859 = 4.3755513$$
$$\text{and } \log. 23743859 = 7.3755513$$
$$\text{or } \log. 237.43859 = 2.3755513$$

In the same way we find the logarithm of 1275.8073. Thus :

log. 1275.8 . . . =3.1057826

$$
\begin{array}{ll}
0.. & 0 \\
7. & 23.8 \\
3 & 1.02 \\
\end{array}
$$

$$\log. 1275.8073 = 3.1057851$$

275.

When we desire to indicate the number corresponding to a given logarithm, we unite num. log : (i. e. *numerus logarithmi.*)

Thus since log. 2=0.3010300

We have inversely :

$$\text{num. log. } 0.3010300 = 2$$

which is read · "the *number* for the logarithm 0.3010300 is 2."

The method of finding the number from the table, when the logarithm is given, is easily deduced from the preceeding rules.

(1) First look for the first three decimals of the mantissa in the column headed o. Then look across horizontally on the same line or on the next lines immediately below, for the last four decimals, in the column headed from o to 9. If these last four are not found here, look one line higher, in which case they must be preceded by an asterisk. If they are not found *exactly*, take the nearest value. We then have in column Num. the first four figures of the required number. We then examine the characteristic of the given logarithm and give our number as many *whole* places as there are units in the characteristic plus 1, and thus place the decimal point.

If the logarithm has a negative characteristic, the number must all be a decimal, and there must be as many zeros placed at the left of the number as there are units in the characteristic less 1, and then the decimal point is placed.

This gives us the number correct to five figures, and the method can readily be understood by the student if he will take his volume of tables and find the logarithms corresponding to the following numbers and vice versa.

log. 22035=4.3431131
num. log. 4.3431131=22035

log. 666.42=2.8237480
num. log. 3.8237480=6664.2

log. 8.7707=0.9430343
num. log. 6.9430343=8770700

log. 0.92904=1.9680344
num. log. 3.9680344=0.0092904

log. 0.051001=2.7075787
num. log. 0.0010411=1.0024

(2) In order to find the additional figures when the exact value of the base decimals of the mantissa cannot be found in the table, we proceed as follows :

First find the first three decimals of the mantissa in column o, as before. Then find in the other columns the *next lower* value to the four last figures, and take the head of the column for the fifth figure, as before.

Then subtract these next lower figures from the last four figures of the given logarithm and look for the nearest value in the corresponding portion of the column of proportional parts. The figure in the P. P. column which corresponds to this difference most closely will be the sixth figure of the number, always taking the next lower value. Then subtract this lower value from the tabular proportion thus taken, and the proportion corresponding to this difference will give the seventh figure. With seven place tables this is as far as the result can be accurately carried, and if the characteristic demands another figure its place must be filled by a zero.

An example will make the procedure clear.

Required : Num. log. 7.3255512.

We find in the table the next smaller logarithm is 7.3755356, hence :

$$\text{given log.} = 7.3755512$$
$$\text{next smaller log.} = 7.3755356$$
$$\text{and its number} = \quad 23743000$$
$$\text{Difference} \quad 156$$

In the column P. P. the next smaller value is 146.4, and the corresponding figure is 8. Then 156—146.4=9·6 which multiplied by 10=96, and the next smaller value is 91.5, for which the number is 5.

Hence : Num. log. 7.3655512=23743850

Also : Num. log. 4.3755512=23743 85

This rule follows evidently from what has been said before. The reasoning is as follows :

The difference between the given logarithm and that for a number smaller by 1 is 183. But the difference between the given logarithm and the next smaller logarithm is 156. Therefore we have the proposition :

$$\text{As} \quad 183:1 \text{ so is } 156:x$$
$$x = \tfrac{156}{183} = 0.85$$

276.

The following examples will answer for practice, and larger numbers will rarely occur in actual examples. The 8 and

9 figure numbers will vary slightly from exactness beyond the seventh decimal, as greater accuracy can hardly be obtained with seven place tables.

The methods of using 5 and 6 figure tables of logarithms will readily be understood from what has been said above, and specific directions for the various editions will be found with them.

EXAMPLES.

(1)   log. 370978=5.5693482

(2)   num. log. 3.8911459=7782.98

(3)   log. 8689836=6.9390116

(4)   num. log. 6.9720151=9375946

(5)   log. 200.36084=2.3018128

(6)   num. log. 0.0692746=1.172937

(7)   log. 0.07787009=2.8913707

(8)   num. log. 3.0911392=0.0012335

(9)   log. 4.501000895=0.6533091

(10) num. log. 0.0901392=1.230664

BOOK XX.

The Applications of Logarithms.

277.

In the following pages will be given examples and rules for the applications of logarithms, and among these, examples will be given of problems which without logarithms could only be solved with greatest difficulty, or in some cases could not be solved at all.

278.

To obtain the product of two numbers we add the logarithms of the factors; the sum is then the logarithm of the product, which can be obtained from the table as explained in § 258, 1. Stated as a general formula with letters for symbols we have :

$$x = abc$$

in which a, b, and c are factors and x is their product :

$$\log. x = \log. a + \log. b + \log. c$$

EXAMPLE. Find x from the following equation :

$$x = 823 \cdot 1305 \cdot \tfrac{3}{7} \cdot (2.40067) \, (0.0067925)$$

We have :

$$
\begin{aligned}
\log. \quad 823 \qquad &= 2.9153998 \\
\log. \quad 1305 \qquad &= 3.1156105 \\
\log. \tfrac{3}{7} = \log. \ 0.4285714 \ &= 1.6320^2{}_{10}^{18} \\
\log. \ 2.40067 \qquad &= 0.380^3{}_{127}^{198} \\
\log. \ 0.0067925 \qquad &= 3.8320296 \\[4pt]
\hline
\log. \ x &= 3.8753956 \\
x &= 7505.776
\end{aligned}
$$

279.

In order to divide one number by another, subtract the logarithm of the divisor from the logarithm of the dividend; the remainder will be the logarithm of the quotient.

If $x=\frac{a}{b}$ in which a is the dividend, b the divisor and x the quotient then we have:

$$\log. x = \log. a - \log. b.$$

EXAMPLE. Find x from the following equation:

$$x = \frac{25.0035}{7123.0409}$$

log. 25.0035=1.3980008
log. 7123.0409=3.8526653

log. x=3.5453355
x=0.003510229

In the subtraction we had to borrow 1, from the units place, and then 0—3, gave 3 for the characteristic.

If either the dividend or divisor, or both, consist of factors, we first find the logarithms of the factors, and add them, and then perform the subtraction. Thus:

$$x = \frac{abc}{de}$$

log. x=log. (abc)—log. (de)
or log. x=log. a+log. b+log. c—(log. d+log. e)
or log. x=log. a+log. b+log. c—log. d—log. e

EXAMPLE. Find x from the following equation:

$$x = \frac{0.035689 \cdot 6.083769}{34.595 \cdot 0.0050602}$$

We have for the numerator:

log. 0.035689=2.5525344
log. 6.083769=0.7841727

1.3367071

for the denominator:

log. 34.595=1.5390133
log. 0.0050602=3.7041677

1.2431810

$$\text{whence}: \quad 1.3367071$$
$$1.2431810$$
$$\text{log. } x = 0.0935261$$
$$x = 1.240298$$

280.

In order to raise a number to a given power, the logarithm of the number is multiplied by the exponent of the power; the product will be the logarithm of the power. (See § 258, 3).

Thus:

log. a^4=log. $aaaa$=log. a+log. a+log. a+log. a whence log. a^4 =4 log. a.

In general:

$$\text{log. } a^n = n \text{ log. } a$$
$$\text{or} \quad \text{log. } a^x = x \text{ log. } a$$

EXAMPLE 1. Find x from the following equation:

$$x = (1.3504)^{22}$$

We have:

$$\text{log. } 1.3504 = 0.1304624$$
$$22$$

$$\overline{}$$
$$2609248$$
$$2609248$$
$$\overline{}$$
$$\text{log. } x = 2.8701728$$
$$x = 741.6052$$

EXAMPLE 2. Find x from the following:

$$x = \left(\tfrac{200}{331}\right)^{10}$$

$$\text{log. } 200 = 2.3010300$$
$$\text{log. } 331 = 2.5198280 \quad (\S\ 272)$$
$$\overline{}$$
$$1.7812020$$
$$10$$
$$\overline{}$$
$$\text{log. } x = 3.8120200$$
$$x = 0.006486643$$

Here we have the positive mantissa

$$0.7812020 \times 10 = 7.8120200$$

and the negative characteristic $1 \times 10 = 10$.

$$\text{hence log. } x = 3.8120200$$

281.

In order to extract any given root we divide the logarithm of the number by the exponent of the root; the quotient will be the logarithm of the root.

Thus (§258,4):

$$\log. \sqrt[5]{a} = \log. a^{\frac{1}{5}} = \tfrac{1}{5}\log. a$$

and in general:

$$\log. \sqrt[n]{a} = \frac{1}{n}\log. a$$

EXAMPLE.

$$x = \sqrt[7]{2}$$

$$\log. 2 = 0.3010300$$

$$7)0.3010300$$

$$\log. x = 0.0430043$$

$$x = 1.104089$$

If we have to extract a root of a proper fraction, i. e., one of which the logarithm has a negative characteristic, we must add as many units to the negative characteristic as will make it positive and will also be exactly divisible by the exponent; also, of course, subtracting the same number of units, in order that the value of the logarithm be not altered. We then divide both quantities by the given exponent, as follows:

EXAMPLE 2. Required:

$$x = \sqrt[5]{0.0375}$$

$$\log. 0.0375 = \overline{2}.5740313$$

adding $-5+5$.

$$5)-5+3.5740313$$

$$-1+0.7148063$$

$$\log. x = \overline{1}.7148063$$

$$x = 0.5185687$$

282.

If we have to multiply two or more factors together, which factors are also powers, we take the logarithms of the roots, and multiply them first by their respective exponents, and then add the results, to effect the multiplication, as will appear by the following examples :

(1) $$x=a^2b^3=aaa.bb$$

log. $x=$log. $a+$log. $a+$log. $a+$log. $b+$log. b

or   log. $x=3$ log. $a+2$ log. b

(2) $$x=a^m b^n c^n$$

log. $x=m$ log. $a+n$ log. $b+n$ log. c

(3) $$x=a^{\frac{m}{n}}\sqrt{b^n}a^{\frac{m}{n}}b^{\frac{n}{m}}$$

log. $x=\dfrac{m}{n}$ log. $a+\dfrac{n}{m}$ log. b

283.

It is evident that we can only obtain a product by means of logarithms when the factors are monomials. When the factors consist of several members, each factor must be considered as a single member, or if the various members are numbers they must be collected into single members before the logarithmic calculations are made. Thus, for example :

(1) $$x=a^m(a+b)^n$$

log. $x=m$ log. $a+n$ log. $(a+b)$

and $a+b$ must be determined before its logarithm can be taken.

(2) $$x=\frac{(a+b^m)^n\ (a+b)}{(a-1)^n}$$

log. $x=n$ log. $(a+b^m)+$log. $(a+b)-n$ log. $(a-1)$ '

(3) $$x=\sqrt[15]{(\tfrac{15}{32}-\sqrt[5]{\tfrac{3}{1144}})}$$

log. $3=0.4771213$

log. $1144=3\ 0584260$

$3\ 4186953$

$-5+5.$

$5)-5+2.4186953$

num. log. $-1+0.4837391=0.3046064$

which subtracted from $\tfrac{15}{32}=0.46875$

gives $\tfrac{15}{32}-\sqrt[5]{\tfrac{3}{1144}}=0\ 1641436$

$$\log. \ 0.1641436 = \overline{1}.2152240$$
$$-15 + 15$$
$$15)\ \overline{-15} + 14.2152240$$
$$\overline{1} + 0.9476816$$
$$\log. \ x = \overline{1}.9476816$$
$$x = 0.886506$$

284.

Since we are able by means of logarithms, readily to extract any root whatever, it is also clear that we may use them to solve all pure equations of the higher degrees. Suppose, for example, that it is required to find the value of x, from the following equation, (§ 219):

$$\tfrac{3}{4}x^{11} + 0.501 = 2x^{11} + 6.05$$

This is a pure equation of the 11th degree.

Collecting the values of x^{11}, we have :

$$\tfrac{3}{4}x^{11} - 2x^{11} = -0.501 + 6.05$$
$$-\tfrac{5}{4}x^{11} = 5.549$$
$$x = \sqrt[11]{-4.16175} \qquad\qquad (\S\ 216,3)$$
$$\log. \ 4.16175 = 0.6192760 \ (n)$$
$$\log. \ x = \frac{0.6192760 \ (n)}{11} = 0.0562978$$
$$x = -1.138407$$

The (n) appended to the logarithm means that the number to which it belongs has the minus sign. Since the logarithms of whole numbers are positive, and those of proper fractions are negative, it is evident that we cannot find in the tables, logarithms for negative numbers. In order therefore to work with logarithms for negative numbers, we consider the numbers as positive, and append the symbol (n) to the logarithms, and give the negative sign to the result.

285.

By the use of logarithms it is often easy to determine the unknown quantity in an equation, when the unknown appears as an exponent. Suppose, for example, it is required to find what power of 2 will equal 64, we have the following equation :

$$2^{x} = 64$$

Taking the logarithms of both sides, we have :

$$x \log 2 = \log 64$$

$$x = \frac{\log 64}{\log 2} = \frac{1.8061800}{0.3010300} = 6$$

The student must be careful not to confuse $\frac{\log 64}{\log 2}$ with $\log.$ $\frac{64}{2} = \log 64 - \log 2$. In the latter case we subtract one logarithm from the other, but in the former case we divide the one by the other. In this latter case if desired we can consider the logarithms as ordinary numbers, taking again the logarithms of each of them from the table and subtracting the logarithm of the divisor from the logarithm of the dividend to obtain the logarithm of the quotient, and thus save the labor of the long division.

In general, when :

$$a^x = b$$

$$x \log a = \log b$$

$$x = \frac{\log b}{\log a}$$

Again if :

$$a^{mx} b^{n-\frac{x}{2}} = c^{k-x} c$$

We have :

$$mx \log a + (n - \tfrac{x}{2}) \log b = (k - x) \log c + \log d$$

$$xm \log a + n \log b - \tfrac{x}{2} \log b = k \log c - x \log c + \log d$$

$$xm \log a - \tfrac{x}{2} \log b + x \log c = k \log c + \log d - n \log b$$

$$x = \frac{k \log c + \log d - n \log b}{m \log a + \log c - \tfrac{1}{2} \log b}$$

286.

EXAMPLE I. Required x, from the equation :

$$a \cdot c^{mx} - b \cdot c^{\frac{mx}{2}} = d$$

Solution. In order to simplify the work let us temporarily represent the $c^{\frac{mx}{2}} = z$, whence $c^{mx} = z^2$, and we have (§ 231)

$$az^2 - bz = d$$

$$z = \frac{b \pm \sqrt{b^2 + 4ad}}{2a}$$

replacing z by its value $c^{\frac{mx}{2}}$, we have :

$$c^{\frac{mx}{2}} = \frac{b \pm \sqrt{b^2 + 4ad}}{2a}$$

$$\frac{mx}{2}\log. c = \log. \left(\frac{b \pm \sqrt{b^2 + 4ad}}{2a}\right)$$

$$x = \frac{2}{m \log. c} \cdot \log. \left(\frac{b \pm \sqrt{b^2 + 4ad}}{2a}\right)$$

EXAMPLE 2. Find x, from the equation :

$$a^x + \frac{1}{a^x} - b = 0$$

Solution. Multiplying by a^x, and then putting $a^x = z$, we readily find, according to the method given in § 231 :

$$x = \frac{\log. \left[\frac{1}{2}(b \pm \sqrt{b^2 - 4})\right]}{\log. a.}$$

287.

EXAMPLE 3. Interpolate four terms between 2 and 8, so that all six terms shall form a geometrical progression.

Solution. Given $a = 2$, $l = 8$, $n = 6$, and e, required.

We have from § 253 :

$$l = ae^{n-1}$$

whence :

$$e = \sqrt[n-1]{\frac{l}{a}}$$

$$\log. e = \frac{\log. l - \log. a}{n - 1}$$

$$\log. e = \frac{\log 8 - \log. 2}{n - 1}$$

$$e = 1.319508$$

The series will therefore be :

$$2 ; \quad 2 \cdot (1.319) ; \quad 2 \cdot (1.319)^2 ; \quad 2 \cdot (1.319)^3 ; \quad 2(1.319)^4 ; \quad 8.$$

288.

EXAMPLE 4. It is related that Sissa, the inventor of chess, was requested by the Hindu Rajah Scheran, to demand a reward for his ingenious invention. Sissa modestly requested that he should have the sum of the number of grains of rice that could be counted by the squares of the chess board, if counted in the following manner. On the first square was to be placed 1 grain, on the second square two grains, on the third square 4 grains, on the fourth square 8 grains, &c., increasing in this progression: $1+2+4+8+16\ldots+2^{63}$; that is doubling on each square the amount on the preceding square until the whole 64 squares had been counted. The rajah was at first offended at a demand upon him so small as to be beneath his dignity, but ordered the gift to be made, when he was surprised to find it far beyond the power of himself or any other monarch. How much would the total be?

Solution. Given $a=1$, $e=2$, $l=2^{63}$, and s required.

We have from § 254 :

$$s = \frac{le-a}{e-1}$$

$$s = \frac{2^{63} \cdot 2 - 1}{2-1} = 2^{64} - 1$$

It will be close enough to assume :

$$s = 2^{64}$$

whence   log. $s = 64$ log. 2

$$s = 18446750000000000000c$$

More exactly, the sum would be

$$s = 18446744073709551615$$

The last figures cannot be found by means of logarithms, as they exceed the limits of any table which has ever been calculated. It is estimated that if the whole surface of the earth was under cultivation it would require the harvests of seventy years to equal this amount.

289.

EXAMPLE 5. A wine butt contains 100 gallons of wine, which we will call a. From this 1 gallon $=b$, is drawn, and an equal amount of water poured in. When the water and wine are thoroughly mixed the amount b is again drawn and replaced with water. After this operation has been repeated 20 times, how much wine remains?

Solution. The problem may be stated in the following manner: If 1 gallon is taken from 100 gallons and replaced by water there must remain 99 gallons of wine, and the mixture will consist of $\frac{1}{100}$·99 of wine and $\frac{1}{100}$ of water. When the second draught is made, there will remain $99-\frac{1}{100}$·99 of wine and $1-\frac{1}{100}$·1 of water. After the third draught there will be $(99-\frac{1}{100}$·99) of pure wine, etc., etc.

In general, therefore, if $a=$ the original amount of wine, and b the amount of each draught, we have taken the proportion $\frac{b}{a}$ of the original amount the first time and left $a-\frac{b}{a}=a-b$ in the the vessel. After putting in the water, we take out the second time $\frac{b}{a}(a-b)$ of wine and there will remain :

$$(a-b)-\frac{b}{a}(a-b)=\frac{a(a-b)-b(a-b)}{a}=\frac{(b-b)^2}{a}.$$

After the third time we have left :

$$\frac{(a-b)^2}{a}=\frac{b}{a}\frac{(a-b)^2}{a}=\frac{(a-b)^3}{a^2}$$

After the fourth time $\frac{(a-b)^4}{a^3}$, etc.

If x be the amount of wine remaining after the nth draught we have in general :

$$x=\frac{(a-b)^n}{a^{n-1}}$$

$$\log. x=n \log. (a-b)-(n-1) \log. a$$

If, as in the example : $a=100$, $b=1$, $n=20$, we have

$$\log. x=20 \log. 99-19 \log. 100$$

$$x=81.79072$$

BOOK XXI.

Compound Interest.

290.

When the interest on any capital is added to the principal to form a new principal upon which interest is again computed for the next period of time, the capital is said to be at Compound Interest.

It is evident that at compound interest the capital must increase very rapidly, the rate of increase itself also increasing. Thus, for instance, if \$100 is at 5 per cent. compound interest, we have at the end of the first year a capital of $100 + \frac{100}{100} \cdot 5 =$ \$105 for the next year; at the end of the second year this has grown to $105 + \frac{105}{100} \cdot 5 = \$110\frac{1}{4}$; at the end of the third year, to $110\frac{1}{4} + \frac{110\frac{1}{4}}{100} \cdot 5 = \$115\frac{61}{80}$. etc.

Cases of this sort constantly occur in commercial affairs, particularly in connection with Insurance, Savings Banks, Annuities, and similar financial operations, and the following methods will be found useful in solving the problems which arise.

291.

PROBLEM. If we indicate the original capital by a, the rate per cent. per year by p, the number of years by n, and the final total amount by A ($=$ Accumulation), it is evident that each one of the four qualities A, a, p, and n, is a determinate function of the other three. We shall therefore first derive the various equations which show the relations existing between these four quantities.

We shall first seek the expression for the function of A, from which by reduction we can then readily derive the others.

At the end of the first year the original capital, *a*, will have become :

$$a+\frac{p}{100}\cdot p=a\left(1+\frac{p}{100}\right)$$

That is, we obtain the total at the end of the first year if we multiply the capital by $1+\frac{p}{100}$. At the beginning of the second year we have the new capital $a\left(1+\frac{p}{100}\right)$ and at the end of the second year this is again to be multiplied by $1+\frac{p}{100}$ so that the amount will be :

$$a\left(1+\frac{p}{100}\right)+\frac{a(1+\frac{p}{100}}{100}\cdot p=a\left(1+\frac{p}{100}\right)\left(1+\frac{p}{100}\right)=a\left(1+\frac{p}{100}\right)^2$$

Likewise at the end of the third year we have :

$$a\left(1+\frac{p}{100}\right)^3,\text{ etc.}$$

and in general, at the end of *n* years the total will be :

$$a\left(1+\frac{p}{100}\right)^n\text{ or }A=a\left(1+\frac{p}{100}\right)^n$$

If, for sake of simplicity we set the expression $1+\frac{p}{100}=z$, we have :

$$A=a\cdot z^n$$

For $p=5\%$, $z=1.05$; for $p=4\%$, $z=1.04$; for $p=3\frac{1}{2}\%$, $z=1.035$, etc., so that z can instantly be given for any rate per cent. Taking the logarithms of both sides of the above equation, we have :

$$\log. A=\log. a+n\log. z \qquad (1)$$

$$\log. a=\log. A-n\log. a \qquad (2)$$

$$\log. z=\frac{\log. A-\log. a}{n} \qquad (3)$$

$$n=\frac{\log. A-\log. a}{\log. z} \qquad (4)$$

292.

EXAMPLE I. What will be the amount of a capital of $6000 at 5% compound interest for 16 years ?

Solution. Given $a=6000$, $n=10$, $z=1.05$, and A required. We have :

$$A=az^n$$

$$\text{log. } z=0.0211893$$
$$16$$

$$\overline{1271358}$$
$$211893$$

$$\overline{0.3390288}$$
$$\text{log. } a= 3.7781513$$

$$\overline{\text{log. } A=4.1171801}$$
$$A=\$13097.25$$

If we subtract the original capital from the final total we can see how much the increase has been. Thus in the above example :

$$13097.25—6000=\$7097.25$$

or more than the original amount.

293.

Example 2. What must be the original capital, which at 4% compound interest, amounts to \$300 in 10 years.

Solution. Here we have : $z=1.04$; $n=10$; $A=300$; and a required.

We have from equation (2) § 292 :

$$\text{log. } a= \text{log. } A - n \text{ log. } z$$

$$\text{log. } z=0.0170333$$
$$10$$

$$\overline{0.1703330}$$
$$\text{log. } A=2.4771213$$

$$\overline{\text{log. } a=2.3067883}$$
$$a=\$202.67$$

This value \$202.67 is the discount value of \$300 at 4% for 10 years.

294.

Example 3. A capital of \$900 increased in 12 years to \$1100; what was the rate of compound interest?

Solution. Given $A=1100$; $a=900$; $n=12$; p required.

We shall first find $z=1+\dfrac{p}{100}$, and can then easily find p.

We have from equation (3) § 291 :

$$\log z=\frac{\log A-\log a}{n}$$

$\log A=3.0413927$

$\log a=2.9542425$

divided by 12)0.0871502

$\log z=0.0072625$

$z=1.01686$, say 1.017

$p=0.017=1\tfrac{7}{10}\%$

295.

EXAMPLE 4. How long a time must a capital of $600 remain at 5% compound interest, in order to amount to $800 ?

Solution. Given $a=600$; $z=1.05$, $A=800$, n required.

We have from equation (4), § 291 :

$$n=\frac{\log A-\log a}{\log z}$$

$\log A=2.9030900$

$\log a=2.7781513$

0.1249387

$\log z=0.0211893$

$$n=\frac{0.1249387}{0.0211893}=5.89 \text{ years.}$$

296.

The following examples should be worked out by the student with the aid of his table of logarithms.

EXAMPLE 1. What would be the *increase* in a capital of $2000 at 2% compound interest for 20 years ?

Answer. $2971.88.

EXAMPLE 2. What is the present value of a capital, which at 5% compound interest would amount to $1000 in 15 years?

Answer. We must discount $1000 for 15 years, that is, find a capital a, which, at 15 years at 5%, would make a total of $1000. From formula (2) we find $a=$481.02.

EXAMPLE 3. A money lender loaned $500 for $2\frac{1}{2}$ years, under agreement that at the expiration of that time the borrower was to pay him $700. What rate of compound interest did this represent?

Answer. Over 14 per cent.

EXAMPLE 4. What would be the amount of a capital of $6000 at 5%, for 10 years, the interest being compounded semi-annually, i. e., $2\frac{1}{2}$% being added to the principal every six months?

Answer. Since the interest is added every six months, the number of times, n, in 10 years would be 20, i. e., $n=20$, and we take $2\frac{1}{2}$% instead of 5%. These values give $A=\$9831.70$.

297.

When a given capital, such as $100, is at interest at a given rate, such as 6%, but the rate is computed every three months instead of yearly, it is not correct to assume that the quarterly rate would be $\frac{6}{4}=1.5$%. If the interest is to be computed and compounded every quarter, it is evidently correct to assume the quarter-year as the unit of time, and hence we can find the correct rate from the formula, $A=az^n$, by making the rate $=x$ and $n=4$, whence:

$$100 \cdot \left(1+\frac{x}{100}\right)^4 = 1.06$$

whence: $1+\dfrac{x}{100}=\sqrt[4]{1.06}=1.01467$

and $x=1.467$% instead of 1.5%.

EXAMPLE. Suppose a capital of $20000 at 5% compound interest, what would be the amount at the end of $12\frac{1}{2}$ years?

Answer. Here $a=20000$, $z=1.05$, $n=12\frac{1}{2}$, and from:

log. A=log. 20000+$12\frac{1}{2}$ log. 1.05

we get: $A=\$36804.10$

If we had taken the amount first for 12 whole years and then $\frac{5}{2}$% for the remaining half-year, the difference would have been about $11.

298.

PROBLEM. It is desired to find the number of years required for a given capital at a given rate of compound interest, to double, triple, etc., itself.

Solution. Let a be the original capital, $z = 1 + \frac{p}{100}$, $m =$ the number of times the capital is to be multiplied, and n the required number of years. We then have for the total amount for any given multiplier m,

$$A = ma$$

whence $az^n = ma$

or dividing both sides by the common factor a :

$$z^n = m$$

$$n \log. z = \log. m$$

$$n = \frac{\log. m}{\log. z}$$

We see therefore that the required time in which a capital a shall be increased m-fold, is a function of the rate per cent., and is entirely independent of the amount of the capital a. The same time is required for one dollar to double itself as would be required for a million dollars.

EXAMPLE. How many years would be required for a given capital to double itself, at 5% compound interest?

Answer. Here $m = 2$, and $z = 1.05$, whence

$$n = \frac{\log. 2}{\log. 1.05} = \frac{0.3010300}{0.0211893} = 14.2 \text{ years.}$$

In the same way we find that for 4%, the time required for doubling is between 17 and 18 years; at 3% it is about 23.4 years.

299.

PROBLEM. Suppose that a capital a, in addition to the compounding of its interest has a certain amount added to it at the end of every year, so that the interest for the following year is also computed upon this additional amount. It is required to find an expression by which the increase can be determined at the end of any given number of years.

Solution. Let a be the original capital, $z = (1 + \frac{p}{100})$, $b =$ the extra amount added at the end of every year, $n =$ the number of years. and $A =$ the required total at the end of n years.

At the end of n years the original capital would have become az^n; the amount added at the end of the first year would have been at interest one year less, or $n - 1$ years, and hence, considered by itself, it would have become $b \cdot z^{n-1}$; likewise the same amount b, added at the end of the second year would have become $b \cdot z^{n-2}$, etc. The addition at the end of the next to the last year would have earned only one year's interest, and the amount added at the end of the last year would have earned no interest at all. Indicating the sum of the various amounts, we have for the total :

$$A = az^n + bz^{n-1} - bz^{n-2} + bz^{n-3} + \ldots + bz + b$$

This series we now desire to collect into one single expression. We see that all the terms except the first one (az^n) form a geometrical series, in which, *reading backward*, b is the first term, z the exponent, and bz^{n-1} is the last member. We then have for the sum, according to § 256 :

$$\frac{bz^{n-1} \cdot z - b}{z-1} = \frac{bz^n - b}{z-1} = \frac{b(z^n - 1)}{z-1}.$$

If we, therefore, put this expression for the sum of the series, we have :

$$A = az^n + \frac{b(z^n - 1)}{z-1} \qquad (1)$$

whence :

$$a = \frac{A}{z^n} + \frac{b(z^n - 1)}{z^n(z-1)} \qquad (2)$$

$$b = \frac{(A - az^n)(z-1)}{z^n - 1} \qquad (3)$$

$$n = \frac{\log. [A(z-1) + b] - \log. [a(z-1) + b]}{\log. z} \qquad (4)$$

The expression for the value of z is not given because it leads to a very complicated equation of a high degree.

300.

If in the above formula (1) the extra amount added each year is equal to the original capital, so that we have $b = a$, the equation can be still further simplified. Putting a for b, we have :

$$A = \frac{a(z^{n+1} - 1)}{z-1} \qquad (1)$$

whence :

$$A = \frac{A(z-1)}{z^{n+1} - 1} \qquad (2)$$

and $n+1 = \dfrac{\log. \; [A(z-1)+a]- \log. \; a}{\log. \; z}$ (3)

301.

If, instead of adding a constant amount b, at the end of each year, a constant amount b is *subtracted*, and a formula is required for the total at the end of n years ; we have under these conditions, at the end of the

    1st year : $az-b$
    2nd year : $(az-b)z-b=az^2-bz-b$
    3rd year : az^3-bz^2-bz-b
    10th year : $az^{10}-bz^9-bz^8-\ldots-bz-b$

or in general at the end of the nth year :

$$A=az^n-(bz^{n-1}+bz^{n-2}+bz^{n-3}+\ldots+bz+b)$$

Again we have a geometrical progression in the parenthesis, and substituting for this the expression for its sum, we have :

$$A=az^n-\frac{b(z^n-1)}{z-1} \qquad (1)$$

$$a=\frac{A}{z^n}+\frac{b(z^n-1)}{z^n(z-1)} \qquad (2)$$

$$b=\frac{(az^n-A)(z-1)}{z^n-1} \qquad (3)$$

$$n=\frac{\log. \; [A(z-1)-b]- \log. \; [a(z-1)-b]}{\log. \; z} \qquad (4)$$

If the amount b, which is withdrawn at the end of each year, be less than the yearly amount of interest on the original capital, the total amount A, will evidently increase every year. If, however, the amount withdrawn be greater than the yearly interest, the total will evidently be smaller every year, and in the course of time will become 0. If the operation is continued still further, the relations of debtor and creditor will be reversed.

302.

If the amount b, which is withdrawn yearly, will exhaust both capital and interest in n years, we must have in formula (1) of the preceding paragraph, $A-0$, whence the two members of the right hand side must be equal to each other, and

$$\frac{b(z^n-1)}{z-1}=az^n \qquad (1)$$

From this equation, when of the four quantities a, b, n and z, any three are given, we can find the fourth, § 299.

We have :
$$a = \frac{b\,(z^n - 1)}{z^n\,(z-1)} \tag{2}$$

$$b = \frac{a\,(z-1)z^n}{z^n - 1} \tag{3}$$

$$n = \frac{\log. b - \log. [b - a\,(z-1)]}{\log. z} \tag{4}$$

303.

EXAMPLE 1. What will be the total amount of a capital of \$5000, at 5% compound interest for 25 years, if \$200 is added to the principal at the end of each of the 25 years?

Solution. Here $a = 5000$, $b = 200$, $z = 1.05$, $n = 25$. We have from § 299 :

$$A = az^n + \frac{b\,(z^n - 1)}{z - 1} *$$

log. $z =$ 0.0211893

       25

    1059465

    423786

log. $z^{25} =$ 0.5297325

   $z^{25} =$ 3.386355

$*(z^{25} - 1) =$ 2.386355

log. $(z^{25} - 1) =$ 0.3777350

log. $b =$ 2.3010300

Sum $=$ 2.6787650

$*$log. $z - 1 =$ 2.6989700

Diff. $=$ 3.9797950

$b\dfrac{(z^n - 1)}{z - 1} =$ 9545.42

log. $z^{25} =$ 0.5297325
log. $a =$ 3.6989700

Sum $=$ 4.2287025

$az^{25} =$ 16931.77

$az^{25} =$ 16931.77

$b\dfrac{(z^n - 1)}{z - 1} =$ 9545.42

$A =$ \$26477.19

304.

EXAMPLE 2. Suppose a capital of \$5500 at $4\frac{1}{2}$% compound interest. What will remain at the end of 30 years, if \$100 is taken out at the end of each year?

*The binomial quantities $(z - 1)$ and $(z^n - 1)$ must be reduced to single quantities before their logarithms can be taken.

Solution. Here $a=5500$, $b=300$, $z=1.045$, $n=30$. We have from § 301:

$$A=az^n-\frac{b\,(z^n-1)}{z-1}$$

log. $z=0.0191163$

$\qquad\qquad 30$

log. $z^{30}=0.5734890$

$\qquad z^{30}=\ \ 3.74532$

$\qquad z^{30}-1=\ \ 2.74532$

log. $(z^{30}-1)=0.4385930$

$\qquad$ log. $b=2.4771213$

$\qquad\qquad\qquad 2.9157143$

log. $(z-1)=2.6532125$

$\qquad\qquad\qquad 4.2625018$

$b\dfrac{(z^{30}-1)}{z-1}=18302.13$

log. $z^{30}=0.5734890$

log. $a=3.7403627$

$\qquad\qquad 4.3138517$

$az^{30}=\ 20599.26$

$b\dfrac{(3^{30}-1)}{z-1}=\ 18302.13$

$A=\ \$2297.13$

305.

EXAMPLE 3. A capital of \$6000 at $2\frac{1}{2}\%$ compound interest for 10 years, has been reduced by yearly withdrawals, to \$1000. How much has been withdrawn each year?

Solution. We have given $A=1000$, $z=1.025$, $a=6000$, $n=10$.

From § 301 :

$$b=(az^n-A)\frac{(z-1)}{(z^n-1)}$$

10 log. $z=0.1072390$

log. $a=3.7781513$

$\qquad\qquad 3.8853903$

$az^{10}=\ 7680.514$

$A=\ 1000.000$

$az^{10}-A=\ 6680.514$

$z^{10}=1.2800859$

$z^{10}-1=0.2800859$

log. $(az^{10}-A)=3.8248100$

log. $(z-1)=2.3979400$

$\qquad\qquad\qquad 2.2227500$

log. $(z^{10}-1)=1.4472913$

$\qquad$ log. $b=2.7754587$

$\qquad\qquad b=\$596.30$

306.

EXAMPLE 4. A capital of $30000 is at 5% compound interest, and $3500 is withdrawn every year. In how many years will the entire amount be withdrawn?

Solution. We have :

$$a = 30000, \quad b = 3500, \quad z = 1.05$$

From § 302 :

$$n = \frac{\log b - \log [b - a(z-1)]}{\log z}$$

$$z - 1 = 0.05$$
$$a = 30000$$
$$\overline{}$$
$$a(z-1) = 1500$$
$$b = 3500$$
$$\overline{}$$
$$b - a(z-1) = 2000$$

$$\log b = 3.5440680$$
$$\log [b - a(z-1)] = 3.3010300$$
$$\overline{}$$
$$0.2430380$$

$$\log z = 0.0211893$$

$$n = \frac{0.2430380}{0.0211893} = 11.47$$

or about $11\frac{1}{2}$ years.

307.

EXAMPLE 5. A man having an annuity of $200 a year for 25 years desires to sell it in order to obtain the principal. Supposing the rate of interest to be $3\frac{3}{4}$%, what is the present value of the annuity?

Here we have required to find a capital, a, which, together with its interest at $3\frac{3}{4}$% for 25 years, would just be exhausted by yearly withdrawals of $200 for the same length of time. We have from § 302 :

$$a = \frac{b(z^n - 1)}{z^n(z-1)}$$

in which $b = 200$, $z = 1.0375$, $n = 25$.

$$\log. \; z = 0.0159881$$

$$25$$

$$799405$$
$$319762$$

$$\log. \; z^{25} = 0.3997025$$

$$z^{25} = 2.510166$$

$$z^{25} - 1 = 1.510166$$

$$\log. \; (z^{25} - 1) = 0.1790247$$

$$\log. \; b = 2.3010300$$

$$2.4800547$$

$$\log. \; z^{25} = 0.3997025 \qquad\qquad 2.4800547$$
$$\log. \; (z - 1) = 2.5470313 \qquad\qquad 2.9737338$$

$$2.9737338 \qquad \log. \; a = 3.5063209$$

$$a = \$3208.64$$

308.

EXAMPLE 6. A man pays \$50 a year during a certain number of years to an insurance association in order to secure for his widow a life pension of \$200 a year. The man died after having paid for 20 years ($=n$) and his widow survived him 8 years ($=m$) during which she received the pension. How much did the association make, or lose; the rate of interest being 4% ($z = 1.04$)?

Solution. We must compute the value of each of the accounts backward independently. The first payment of \$50 made by the man was bearing interest for $n+m$ years, and its final value was therefore az^{n+m}. The second payment was at interest for one year less, and hence its value $= az^{n+m-1}$, etc. The last payment was at interest $m+1$ years, and its value $= az^{m+1}$. In like manner the value b, of the first year's pension after m years was $= bz^m$, and the value of the last year's pension $= bz$.

We therefore have for the total value the sum of the first series less the sum of the second, or :

$$A = az^{n+m} + az^{n+m-1} + az^{n+m-2} + \ldots az^{m+1} - (bz^m + bz^{m+1} + \ldots + bz)$$

Summing up both series we have :

$$A = \frac{az^{n+m} \cdot z - az^{m+1}}{z-1} - \frac{bz^m \cdot z - bz}{z-1}$$

$$A = \frac{az^{m+1}(z^n-1)-bz(z^m-1)}{z-1}$$

The first number of the numerator might have been found also in the following manner : The amount of a capital composed of n payments of a, at the beginning of each year would be, at the end of n years equal to $\frac{az(z^n-1)}{z-1}$. Now, if this capital stands for m years more, it would become $= \frac{az^{m+1}(z^n-1)}{z-1}$

The formula may also be written :

$$A = [az^m(z^n-1)-b(z^m-1)]\frac{z}{z-1}$$

In the given example, $a=50$, $b=200$, $n=20$, $m=8$, and $z=1.04$, and these values give in the example a gain for the insurance association of $202.

It is also easy to derive a formula for the case in which the payments are made semi-annually, or one for the case in which the premiums were paid at the beginning of the year and the pension payments at the end of the year, etc.

Such a great variety of cases occur in this branch of mathematics, bearing on legal, political and commercial matters, that only the fundamental principles can here be shown. We shall therefore close this book with two more difficult problems for those who desire to go more deeply into the subject.

309.

PROBLEM. What is the present value a, of a series of yearly payments which form a geometrical progression of which the first member $=b$, and the exponent $=e$: that is, the first payment $=b$, the second $=be$, and the third $=be^2$, and the nth $=be^{n-1}$? The payments to increase or diminish in the ratio $1:e$, in which e may be greater or less than unity.

Solution. Let z be $=1+\frac{p}{100}$, p being the rate of interest.

At the end of the nth year the value of the

1st payment $=bz^{n-1}$
2d payment $=be \cdot z^{n-2}$

$$n-1 \quad \text{payment} = be^{n-1} \cdot z$$
$$n \quad \text{payment} = be^{n-1}$$

The total value of the sum of all the payments from the beginning of the first year for n years, will be equal to az^n. Whence we must have:

$$az^n = bz^{n-1} + be \cdot z^{n-2} + be^2 \cdot z^{n-3} + \ldots + be^{n-2} \cdot z + be^{n-1}$$

The right hand side of this equation forms a geometrical series of which bz^{n-1} is the first term, be^{n-1} the last term, and $\dfrac{e}{z}$ the exponent, and hence, according to § 256 the sum will be equal to :

$$\frac{be^{n-1}\dfrac{e}{z} - bz^{n-1}}{\dfrac{e}{z} - 1} = \frac{be^n - bz^n}{e - z}$$

We therefore have :

$$az = \frac{be^n - bz^n}{e - z}$$

and

$$a = \frac{b(e^n - z^n)}{z^n(e - z)} = \frac{\left[\left(\dfrac{e}{z}\right)^n - 1\right]}{e - z}$$

310.

PROBLEM. What is the present value a, of a series of yearly payments forming the arithmetical progression b, $2b$, $3b$, $4b \ldots nb$, for each year being greater than the preceding by b?

Solution. Let a be the present value, and $z = 1 + \dfrac{p}{100}$, as before, and we have :

$$az^n = bz^{n-1} + 2bz^{n-2} + 3bz^{n-3} + \ldots + (n-1)bz + nb$$

The right side of this equation consists of a combination of arithmetical and geometrical progressions, which may be separated into as many geometrical progressions as there are members, and these summed up separately. We can divide each term into as many parts as there are units in its numerical coefficient, thus:

$$2bz^{n-2} = bz^{n-2} + bz^{n-2}; \quad 3bz^{n-3} = bz^{n-3} + bz^{n-3} + bz^{n-3}, \text{ etc., whence :}$$

$$az^n = \begin{cases} bz^{n-1}+bz^{n-2}+bz^{n-3}+bz^{n-4}+\ldots+bz^2+bz+b & (1) \\ bz^{n-2}+bz^{n-3}+bz^{n-4}+\ldots+bz^2+bz+b & (2) \\ bz^{n-3}+bz^{n-4}+\ldots+bz^2+bz+b & (3) \\ bz^{n-4}+\ldots+bz^2+bz+b & (4) \\ +\ldots+bz^2+bz+b & (5) \\ +\ldots+bz^2+bz+b & (6) \\ +\ldots\ldots\ldots\ldots \\ +\ldots\ldots\ldots \\ +bz^2+bz+b & (n-2) \\ +bz+b & (n-1) \\ +b & (n) \end{cases}$$

Each horizontal line forms a geometrical series, of which the exponent$=z$, and the sums of the series are :

$$(1)=\frac{bz^{n-1}\cdot z-b}{z-1}=b(z^n-1)$$

$$(2)=\frac{bz^{n-2}\cdot z-b}{z-1}=\frac{b(z^{n-1}-1)}{z-1}$$

$$(3)=\frac{bz^{n-3}\cdot z-b}{z-1}=\frac{b(z^{n-2}-1)}{z-1}$$

$$(n-2)=bz^2+bz+b=\frac{bz^2\cdot z-b}{z-1}=\frac{b(z^3-1)}{z-1}$$

$$(n-1)=bz+b=\frac{bz\cdot z-b}{z-1}=\frac{b(z^2-1)}{z-1} \qquad (\S 91, 2)$$

$$(n)=b=\frac{b\cdot z-b}{z-1}=\frac{b(z-1)}{z-1}$$

Placing in the above equation, instead of the various series, their respective sums we have :

$$az^n=\frac{b}{z-1}(z^n-1)+\frac{b}{z-1}(z^{n-1}-1)+\ldots+\frac{b}{z-1}z^2-1)+\frac{b}{z-1}(z-1)$$

$$az^n=\frac{b}{z-1}\left(z^n-1+z^{n-1}-1+\ldots+z^2-1+z-1\right)$$

$$az^n=\frac{b}{z-1}\left(z^n+z^{n-1}+\ldots+z^2+z-(1+1+1\ldots+1)\right)$$

The sum of the series $(1+1+1+\ldots+1)$ in the parenthesis, is equal to n, because there are n members, and the sum of the other geometrical series is :

$$\frac{z^n \cdot z - z}{z - 1} = \frac{z(z^n - 1)}{z - 1}$$

Substituting these values, we have :

$$az^n = \frac{b}{z-1}\left[\frac{z}{z-1}(z^n-1)-n\right]$$

whence

$$a = \frac{b}{z^n(z-1)}\left[\frac{(z^n-1)z}{z-1}-n\right]$$

In computing compound interest for long periods of time it is desirable to have log. z to more than seven decimals, otherwise there may arise errors in the last figures, due to the multiplication of the final error. The following table gives the logarithms of z to 10 decimals for all values of z which are likely to occur in practice :

The logarithm of z should be taken from the table below, and multiplied by n, to give n log. z, and then seven places taken for use with the regular seven place logarithms in the rest of the calculation. If the eighth figure is greater than 5, the seventh figure should be increased by unity. By using these logarithms the final result may almost invariably be depended upon to the last cent.

10 Place Logarithms for z.

z	LOGARITHM	z	LOGARITHM
1.0025	.00108 43813	1.0425	.01807 60636
1.0050	.00216 60618	1.0450	.01911 62904
1.0075	.00324 50548	1.0475	.02015 40316
1.0100	.00432 13738	1.0500	.02118 92991
1.0125	.00539 50319	1.0525	.02222 21045
1.0150	.00646 60422	1.0550	.02325 24596
1.0175	.00753 44179	1.0575	.02428 03760
1.0200	.00860 01718	1.0600	.02530 58653
1.0225	.00966 33167	1.0625	.02632 89387
1.0250	.01072 38654	1.0650	.02734 96078
1.0275	.01178 18305	1.0675	.02836 78837
1.0300	.01283 72247	1.0700	.02938 37777
1.0325	.01389 00603	1.0725	.03039 73009
1.0350	.01494 03498	1.0750	.03140 84643
1.0375	.01598 81054	1.0775	.03241 72788
1.0400	.01703 33393	1.0800	.03342 37555

APPENDIX.

311.

Note to §1. The almost universal decimal system evidently had its origin in the primitive method of counting on the ten fingers. This is, however, not absolutely certain, since in some instances other bases have been found in use by savage nations. Thus if the primitive capacity for counting only extended originally to *four*, the next number would be named "four and one," then "four and two, four and three," etc. Aristotle tells of a people among the ancient Thracians, who counted in this manner, and similar instances have been found by modern anthropologists to exist among savage tribes at the present day. A tribe in the vicinity of Senegal, called the Jalos, have *five* as the base of their system of counting, thus:

ben	*niard*	*niet*	*guyanet*	*guiron*	*guiron ben*
(one)	(two)	(three)	(four)	(five)	(five-one)

Accounts of many other systems used by the Hebrews, Greeks, Romans and other nations will be found in Montucla's *Histoire des Mathematiques*, Vol. I, pp. 45 and 375, also in Klügel's *Mathem. Worterbuch*, 5 *Theil.* p. 1166. A very interesting article on "Primitive Number Systems" by Levi L. Conant, Ph. D., is found in the Smithsonian Report for 1892, p. 583-594, giving also many references to other works.

312.

Note to §6. It is evident that any number may be chosen as the base of a number system, and that simple expressions involving no more characters than there are units in the base of the system, can then be formed. If, for instance, *four* be chosen as the base of a number system, then the first rank of numbers would contain four units instead of ten, and the second rank, sixteen units instead of one hundred, etc. We should then have only the four figures 1, 2, 3, 0 ; and with these, forming the so-called *Tetradik*, all possible numbers could be expressed. In this system the unit is four times greater than the one on its right, and since four units of any rank make one of the next higher rank, we must indicate four units, or one of the first rank, by 10 ; five by 11 ; six by 12 ; seven by 13, and eight, being *two* units of the first rank, by 20; nine $=21$; eleven $=23$; twelve $=30$; fifteen $=33$; sixteen, as four units of the first rank, as one unit of the second rank or $=100$; seventeen $=101$, etc., etc.

In the same way we may take two as the base number, and then have the very simple rule that two units of any rank form one of the next higher rank, when we should be able to indicate all numbers by the two symbols 1 and 0 ; thus, one $=1$; two, being a unit of the next rank, $=10$; three $=11$; four being two units of the first rank, $=100$; five $=110$; seven $=111$; eight, being two units of the second rank, $=$ one of the third rank $= 1000$; nine $=1001$; ten $=1010$, etc. This system was formerly used by the Chinese, except that instead of a 0, they used a short dash (—).

From the practical use of this system to exhibit the peculiar properties of numbers, Leibnitz sought to deduce a basis for creation, by means of which the Jesuit Grimaldi might convert the Chinese to Christianity.

In like manner we may use the number *twelve* as a base, and in this case it is necessary to have two simple symbols for ten and eleven, such for example as α and β. We then write, nine$=9$; ten$=\alpha$; eleven$=\beta$; and twelve being the unit of the first rank$= 10$: thirteen$=11$; twenty-three$=1\beta$; twenty-four$=20$, &c. At one time this system, the *Duodekadik*, was seriously considered instead of the existing, or *Dekadik* ; originally on the curious ground that because there were twelve apostles, there should be twelve numbers. More recently this system has been urged for the practical reason that twelve contains a greater number of factors than ten, which would simplify reckoning to some extent. This reasoning did not seem to offer sufficient advantage to compensate for the inconvenience of abandoning the existing system, although the dozen$=12$, and the gross$=12\cdot12=144$ are firmly fixed in many branches of trade.

We see, therefore, that an endless variety of number systems might be suggested, and it is really a matter of surprise that the ancient Greeks, especially such a brilliant mathematician as Archimedes, never thought of any such number system.

In order to compare the various systems, they are here given in the form of a diagram :

Dyadik	Triadik	Tetradik	Pentadik	Hexadik	Heptadik	Octadik	Enneadik	Dekadik	Enneadekadik	Duodekadik
1	1	1	1	1	1	1	1	1	1	1
10	2	2	2	2	2	2	2	2	2	2
11	10	3	3	3	3	3	3	3	3	3
100	11	10	4	4	4	4	4	4	4	4
101	12	11	10	5	5	5	5	5	5	5
110	20	12	11	10	6	6	6	6	6	6
111	21	13	12	11	10	7	7	7	7	7
1000	22	20	13	12	11	10	8	8	8	8
1001	100	21	14	13	12	11	10	9	9	9
1010	101	22	20	14	13	12	11	10	α	α
1011	102	23	21	15	14	13	12	11	10	β
1100	110	30	22	20	15	14	13	12	11	10

The question may be asked, as to which of these systems is the best. This is largely a matter of custom. Any system may be made practical, and the claims of the *Dekadik*, our existing decimal system, are due to its universal use and consequent advantages of custom. It is only the unfamiliarity which prevents us from being able to write any of the systems readily, and use them in reckoning.

One thing is evident, the larger the base of the number system, the more difficult and larger.is its multiplication table, but also the more rapid are the calculations. The Chinese *Dyadik* system required no multiplication table at all.

It is quite easy to convert a number, expressed in one system, into the equivalent expression in any other system. For example, suppose we have the number 210232 in the tetradik system (base=4), to be converted into our decimal system, we have only to multiply each figure by 4 as many times as indicated by the rank in which it stands. The first figure at the right is in the oth rank, and hence contains the exact number of units indicated ; the second figure is in the first rank, and hence every unit equals 4 of the decimal system ; the third figure is of the second rank, and hence contains 4·4=16 units, &c. Hence the number 210232 in the tetradik system =2350 in the common or decimal system, as follows :

$$
\underset{2}{\underbrace{1024}}\;\underset{1}{\underbrace{256}}\;\underset{0}{\underbrace{64}}\;\underset{2}{\underbrace{16}}\;\underset{3}{\underbrace{4}}\;\underset{2}{\underbrace{1}} = \left\{
\begin{array}{ll}
2= & 2 \\
3\cdot4= & 12 \\
2\cdot16= & 32 \\
0\cdot64= & 0 \\
1\cdot256= & 256 \\
2\cdot1024= & 2048 \\
\hline
& 2350
\end{array}
\right.
$$

In the same manner we find the number 101011 in the dyadik system (base=2) is equal to 43 in the common system. Thus :

$$
\underset{1}{\underbrace{32}}\;\underset{0}{\underbrace{16}}\;\underset{1}{\underbrace{8}}\;\underset{0}{\underbrace{4}}\;\underset{1}{\underbrace{2}}\;\underset{1}{\underbrace{1}} = \left\{
\begin{array}{ll}
1= & 1 \\
1\cdot2= & 2 \\
0\cdot4= & 0 \\
1\cdot8= & 8 \\
0\cdot16= & 0 \\
1\cdot32= & 32 \\
\hline
& 43
\end{array}
\right.
$$

If, on the contrary, it is desired to convert a number from the common system into another, such, for example as the number 2350 into its equivalent expression in the tetradik system, we must divide by the base number 4.

The first quotient gives the number of units in the first rank, and the first remainder, the number of units in the oth rank ;

the second quotient gives the number of units of the second rank, and the second remainder belongs to the first rank, etc. Thus, for example :

$$
\begin{array}{cccccc}
2350 & \overset{\smile}{587} & \overset{\smile}{146} & \overset{\smile}{36} & \overset{\smile}{9} & \overset{\smile}{2} \\
\hline
2 & 3 & 2 & 0 & 1 & 2
\end{array}
$$

and 2350 (dekadik)$=$210232 (tetradik).

Also :

$$
\begin{array}{cccccc}
43 & \overset{2}{21} & 10 & 5 & 2 & 1 \\
\hline
1 & 1 & 0 & 1 & 0 & 1
\end{array}
$$

and 43 (dekadik)$=$101011 (dyadik)

The mechanism of the four fundamental rules is alike in all systems. In addition, in the tetradik system, for example, we carry one unit to the next rank for every four units in the next lower rank, etc., as shown below :

Addition.

	Dyadik.	Tetradik.	Duodekadik.
	10101001	301202	$167\beta89\alpha$
	1111001	133112	$\beta292\alpha1$
Sum :	101100010	1100320	$25\alpha8\beta7\beta$

Subtraction.

	Dyadik.	Tetradik.	Duodekakik.
	100100010	1100230	$25\alpha8\beta7\beta$
	1111001	301202	$\beta292\alpha1$
Remainder :	10101001	133112	$167\beta89\ \alpha$

Multiplication.

	Dyadik.	Tetradik.	Duodekadik.
	10111	2302	$9\beta0\alpha$
	1011	213	$4\beta3$
	10111	20112	25926
	10111	2302	91192
	101110	11210	33834
	11111101	1230132	$40\beta9\alpha46$

Division.

Dyadik.	*Tetradik.*

```
1011) 11111101 (1011        2302) 1230132 (213
      10111                       11210
                                  ─────
      100010                      10313
      10111                       2302

      10111                       20112
      10111                       20112
```

When a number contains only two different figures, it is easy to see at a glance by what numbers (composed of the same two figures) it is divisible. Thus, for example, we see at once that the number 95959595 is divisible by 95 and by 9595.

In the dyadik system, all numbers (including the remainder after a subtraction), are composed of the two symbols 1 and 0, whence Leibnitz based on this system his theory of the divisibility of numbers. Thus, for example, the number 100100100100, is divisible by 100, 1001, 10, 11, etc. If, therefore, we transform a number into the dyadik system we can readily find its factors, and then transform these back into the common system. For example, the number 31393, written in the dyadik system $=111101101000001$, and we see at once that it is divisible by 111 $(=7$; by 1101 $(=13)$; by 10001 $(=17)$; and by 10011 $(=19)$. (See Lambert's *Mathe. Schriften*).

313.

NOTE TO § 10. The statement that the order of the two factors of a product may be chosen at will is easily proved in the following manner : It is undoubtedly just the same whether we take the factor b, a times, or take each unit in b, a times. It is also evidently just the same, when we take unity a times, or take a one time, that is : $a \cdot 1 = 1 \cdot a = a$. If now we take the number of units in b, a times, we must have taken a itself b times. In other words, if we separate b into its units :

$$a\, b = a(1+1+1+1+1\ldots) = a+a+a+a+a+\ldots = b \cdot a.$$
Thus : $4 \cdot 5 = 4(1+1+1+1+1) = 4+4+4+4+4 = 5 \cdot 4$

This law is perfectly general, and may be extended to any number of factors, since the product of two factors can be extended to three ; from three to four, etc. Thus we have : $a \cdot bc = a \cdot cb = bc \cdot a = cb \cdot a$, if in the proceeding demonstration, we take first the two factors a and b, and then the two factors bc and a But we have $ab \cdot c = a \cdot bc$, for it is certainly just the same whether we multiply a and b, and then ab by c, or b by c, and then bc by a. Likewise we have $bc \cdot a = b \cdot ca$, and finally : $abc = acb = cab = cba = bca = bca$, etc.

314.

NOTE TO § 24. The numbers 2 and 5 are contained in 10 without a remainder ; likewise $2 \cdot 2 = 4$, and $5 \cdot 5 = 25$ are contained in $10 \cdot 10 = 1000$.

In order now to prove that a number such as 276, for example, is exactly divisible by 2 because its last figure is so divisible, we must consider the number to be divided into two parts thus : $270 + 6 = 27 \cdot 10 + 6$. We have always the first part divisible by 2 because it is always a multiple of 10, while the last figure is divisible by 2 from the given condition, therefore, the sum must be divisible by 2, thus : $\frac{276}{2} = \frac{270 + 6}{2} = \frac{27 \cdot 10}{2} + \frac{6}{2} = 27 \cdot 5 + 3$. In the same way we have it for 5, and also for $2 \cdot 2 = 4$; $2 \cdot 2 \cdot 2 = 8$, etc., when the last two, or last three figures are divisible by 2 or 5. Thus, for example, in the case of the number 1484, it is divisible by 4 because the last two figures are divisible by 4, thus :

$$\frac{1484}{4} = \frac{1400 + 84}{4} = \frac{14 \cdot 100}{4} + \frac{84}{4}$$

or, 675 is divisible by 5, because its last figure is divisible by 5, thus :

$$\frac{675}{5} = \frac{670 + 5}{5} = \frac{67 \cdot 10}{5} + \frac{84}{4}$$

Any number may be separated into the following parts : Each figure multiplied by as many nines as there are figures following, plus the sum of all the figures, thus :

$$6453 = 6 \cdot 999 + 4 \cdot 99 + 5 \cdot 9 + (6 + 4 + 5 + 3)$$
$$= 6 \cdot 999 + 4 \cdot 99 + 5 \cdot 9 + 18$$

because we have :

$$6453 = 6000 + 400 + 50 + 3 = 6 \cdot 1000 + 4 \cdot 100 + 5 \cdot 10 + 3$$

Now since $1000 = 999 + 1$, $100 = 99 + 1$, etc., we have also :

$$6000 = 6(999 + 1) = 6 \cdot 999 + 6$$
$$400 = 4(99 + 1) = 4 \cdot 99 + 4$$
$$50 = 5(9 + 1) = 5 \cdot 9 + 5$$
$$3 = 3 = 3$$

$$6453 = 6 \cdot 999 + 4 \cdot 99 + 5 \cdot 9 + 18$$

If, therefore, the sum of the figures forming the number 6453, is divisible by 3 or by 9, all the members of the right hand side of the above equation are so divisible, and therefore the number itself is divisible by 3 or 9, thus :

$$\frac{6453}{9} = \frac{6\cdot999 + 4\cdot99 + 5\cdot9 + (6+4+5+3)}{9}$$

315.

NOTE TO § 29. According to the rule given in § 29, we find the greatest common factor of 72 and 168 to be 24.

That the number 24, found by this rule is really a common factor, and also the greatest common factor of these two numbers, is readily seen if we represent the numbers 72 and 168 by lines. Taking any suitable length as a unit, then a line 72 times as long (*a–b*) will represent the number 72, and a line 168 times as long (*c–g*) will represent the number 168.

$$a \overset{72}{\rule{1.5cm}{0.4pt}} b \qquad h \overset{36}{\rule{0.8cm}{0.4pt}} k \qquad c \overset{72}{\rule{1.5cm}{0.4pt}} d \overset{72}{\rule{1.5cm}{0.4pt}} e \rule{0.4cm}{0.4pt} g$$

If now we set off *ab* twice along *cg*, the remainder *eg*=24, however many times it may go into *ab* without a remainder, must also divide exactly into *cd*, *de* and *eg* (the latter being equal to itself). Any greater length, such as *hk*—36, which may divide into *ab*, and therefore into *cd* and *de*, cannot divide the shorter line *eg*, and therefore cannot divide the whole line *cg*=168 without a remainder.

The same is evidently true no matter how many times the smaller number is contained in the greater.

316.

NOTE TO §§ 28, 52, 182.

THEOREM. If *p* is a prime number and *a* and *b* any two whole numbers, neither one of which is divisible by *p* without a remainder, then will their product *a·b* not be divisible exactly by *p*.

Proof 1. If a number *a* is divisible by a prime *p* without a remainder, then when we decompose *a* into its prime factors, *p* will be found among them.

If, on the contrary, two (or more) numbers, *a*, *b*, taken separately, are not divisible by a prime number *p*, it cannot be contained in their product *a·b*, because neither of these containing it among their prime factors can bring it as a factor into their product. It also follows that because a composite number is equal to the product of all its prime factors, that a number a cannot be separated into various sets of prime numbers.

Proof 2. If both factors a and b are greater than p, divide one of them, such as b, by p, calling the quotient $=m$, and the remainder $=r$, this latter necessarily being smaller than p. We then have :

$$\frac{b}{p}=m+\frac{r}{p} \text{ or } b=mp+r$$

Multiplying both sides by $\frac{a}{p}$ we get :

$$\frac{ab}{p}=ma+\frac{ar}{p} \tag{1}$$

If now ab were divisible by p, $\frac{ab}{p}$ would be a whole number and, since ma is a whole number, this would require $\frac{ar}{p}$ to be a whole number also, that is, exactly divisible by p. In order to show the impossibility of this, divide p by r, calling the quotient $=m'$ and the remainder $=r'$, giving :

$$p=m'r+r'$$

Multiplying this by $\frac{a}{p}$ we have :

$$a=m'\frac{ar}{p}+\frac{ar'}{p}$$

If, now, ar were divisible by p and $\frac{m'ar}{p}$, a whole number, then, also, $\frac{ar'}{p}$ must be a whole number, since the total right side of the equation is equal to the whole number a.

But that ar' is not exactly divisible by p, may be shown by dividing r' by p, calling the quotient $=m''$, and the remainder $=r''$. By continuing this operation successively the remainders r, r', r'', r''', etc., must grow smaller and smaller, and finally $=1$.

If, therefore, ab were divisible exactly by p, then, also, ar, ar', ar'', ar''' ...$a\cdot1$ be divisible by p, but this last is equal to a itself, which, by the conditions of the theorem, is *not* exactly divisible by p.

If none of three factors $a\cdot b\cdot c$ is divisible by a prime number p, then the product abc is not divisible by p. This follows from the foregoing, for we have $a \cdot b$ not divisible by p, and if we put $A = ab$, then Ac is not divisible by p, nor is $Aa = abc$, etc. This theorem is a most important one in arithmetic and leads to many others.

317.

NOTE TO § 32. According to the preceding paragraph there is no prime number other than 5 or 2 (such as 3, 7, etc.), nor any multiple thereof (such as $2 \cdot 3$; $4 \cdot 7$; etc.), or in short, no number which cannot be separated into factors of 2 and 5, which will divide 10 ; $10 \cdot 10 = 100$; $10 \cdot 10 \cdot 10. = 1000$, etc., without a remainder. It therefore follows that no fraction of which the numerator and denominator are prime to each other, can be exactly expressed by a decimal fraction if its denominator does not contain the factors 2 and 5. It is also evident that in such cases the decimal must be periodical, i. e., the same figures must return repeatedly in the same order.

If, for example, we convert $\frac{1}{7}$ into a decimal by dividing 7 into 1.000..., it is evident that since none of the remainders can be as great as 7, they must be some of the figures 1, 2, 3, 4, 5, 6, and at the most there can only be six different kinds of remainders. Therefore, necessarily, some one of them must repeat itself, and the decimal then also be repeated.

Thus for example :

$$\tfrac{1}{7} = 0.142857142$$
$$\tfrac{3}{7} = 3 \cdot \tfrac{1}{7} = 3(0.14285714) = 0.42857142$$

We see, therefore, at least, that the greater possible number of places in the repeating decimal must be one less than there are units in the denominator. In order, however, to determine the *exact* number of places of a repeating decimal we must apply the principle referred to in § 166. Thus, for example, every fraction of the form $\dfrac{1}{10^n - 1}$, contains n periodical decimals, as follows :

$$\frac{1}{10^1 - 1} = \frac{1}{9} = 0.111\ldots; \quad \frac{1}{10^2 - 1} = \frac{1}{99} = 0.0101, \text{ etc.}$$

318.

When a periodical decimal is given, it is easy to find the corresponding vulgar fraction. We put $s =$ the periodical fraction, and multiply the equation by 10, or 100, or 1000, choosing such multiplier as will make a whole number *plus* the periodical fraction. We then subtract the first equation from the second, and thus obtain a value for s without decimals, as below :

Let $s = 0.1515...$

Multiply by 100 :

$$100s = 15.1515$$
$$s = 0.1515$$

$$99s = 15$$
$$s = \tfrac{15}{99} = \tfrac{5}{33}$$

Again let :

$$s = 0.321321$$
$$1000s = 321.321321$$
$$s = 0.321321$$

$$999s = 321$$
$$s = \tfrac{321}{999} = \tfrac{107}{333}$$

If the repeating period does not begin immediately after the decimal point, we first multiply by such number as will bring the expression to the desired form, and then proceed as above.

Thus we have :

$$s = 0.25300300...$$
$$100s = 25.300300...$$
$$100000s = 25300.300300...$$
$$100s = 25.300300...$$

$$99900s = 25275...$$

$$s = \frac{25275...}{99900...} = \frac{337}{1332}$$

Again :

$$s = 2.64242...$$
$$10s = 26.4242...$$
$$1000s = 2642.4242...$$
$$10s = 26.4242...$$

$$990s = 2616$$

$$s = \frac{2616}{990} = \frac{436}{165}$$

319.

NOTE TO § 182. There is no power of a vulgar fraction, reduced to its lowest terms (such as $\tfrac{2}{1}$) which will equal an exact whole number. For, if we separate the denominator into its prime factors, such as $\frac{9}{4}=\frac{9}{2\cdot2}$, we see that not only the numerator 9, but also no power of 9, such as $9\cdot9\cdot9\cdot9$, etc., (see § 316) is divisible exactly by 2 ; therefore, it cannot be exactly divisible by any multiple of 2, such as $2\cdot2=4$.

We also have the converse. If the nth root of a whole number N, cannot be expressed in an exact whole number, neither can it be expressed *exactly* by any fraction. For if such a fraction $\frac{a}{b}$ were possible, we should have :

$$\sqrt[n]{N}=\frac{a}{b} \; ; \text{ and } \left(\frac{a}{b}\right)^n=N$$

that is, the fraction $\frac{a}{b}$ multiplied by itself n times, would equal the exact whole number N; which according to § 316, is impossible. The decimals of an irrational root must therefore extend indefinitely, without periods of repetition.

320.

The Theory of Positive and Negative Quantities.

NOTE TO §§ 74–79. It must first be noted that the theory of equations is based upon the idea of opposed magnitudes, and that the rules by which equations are solved must be founded upon the applications of equations to actual cases. The theory of positive and negative quantities, although first practically used in the solution of equations, should really be considered separately, being properly independent of the equations, and actually dependent upon the original conditions of the problem.

From the nature of the case this theory must be abstract in the highest degree, and apparently artificial. An entirely clear understanding of the subject can only come after many practical applications and explanations. This note, therefore, can hardly be made thoroughly intelligible to beginners, but at the same time is really necessary for subsequent study. The correctness of the rules here given follows of itself, and agrees fully with what has already been said in connection with the subject of equations, to which reference is also made.|

(1) Addition. The idea of addition must be considered as consisting of the collection of several separate parts into a

whole; in which act of collection we must take into account the agreement and the opposition of the various parts. It will then be clear that if all the parts are in agreement with each other the sum will be of the same *sense*, so to speak, as are the separate parts and this will be indicated by their having the same *sign*. If, however, the various parts are *opposed* to each other we must first take the sum of the positive parts, and also the sum of the negative parts, and then the smaller of these two sums will cancel an equal value of the opposing sum, and the balance will remain, representing the *sense* of the greater sum, and prefixed by its corresponding sign.

Although this magnitude, in a narrow sense, is a remainder, yet as the result of the collection of a number of parts into a whole, is properly a *sum*, or more precisely an *algebraic sum*, and although the collection of the parts involves an actual subtraction, yet the total performance is really an algebraic addition, as must be admitted, when the subject is broadly considered in its most general form. We therefore see, also, that the axiom that a whole is greater than any of its parts, applies only to magnitudes of the same sign.*

(2) Subtraction. In many cases, especially in the statement of a problem, or the formation of an equation, it occurs that one quantity is to be subtracted from another of opposite sign; or the subtrahend is greater than the minuend, so that the act of subtraction, in its narrow sense of taking one quantity away from another, cannot actually be performed. In order to permit the calculation to proceed, and enable the succeeding operations to be correctly performed, we must consider the subject in a broader light, and hence think of the more general definition of subtraction, as follows : *Subtraction consists in finding that magnitude by which the subtrahend differs from the minuend;* i. e., that quantity which united to the subtrahend will give the minuend. This quantity is then the *real* difference, or in other words the *algebraic difference*, obtained, as we have already seen, by changing the sign of the subtrahend and then taking the algebraic sum with the minuend. These easily remembered facts are entirely general, as will be seen by

*A popular statement of the above may be found in the settling of an estate, in which the *sum* total of the property is found by taking the difference of the assets and liabilities.

xamining the following examples, which include every possible case :

Minuend	$+8$	-8	-8	$+8$	$+2$	-2	-2	$+2$
Subtrahend	$+2$	-2	$+2$	-2	$+8$	-8	$+8$	-8
	$(-)$	$(+)$	$(-)$	$(+)$	$(-)$	$(+)$	$(-)$	$(+)$
Difference	$+6$	-6	-10	$+10$	-6	$+6$	-10	$+10$

It is evident here that the resulting differences (notwithstanding the fact that the signs of the subtrahends have been changed) will give the minuends again when they are added to their respective subtrahends.

It is, therefore, exactly the same, whether a quantity is subtracted, or whether its sign is changed and it is then added.

By the help of equations the truth of this rule can be made apparent.

If we consider the subtrahend combined with the minuend twice—once as it is, and once with the sign changed—the *value* of the minuend will evidently remain unchanged, and an equation can be stated, thus :

$$\text{Minuend.} \quad +8 = +8 - 2 + 2$$
$$\text{Subtrahend.} \quad -2 = \quad -2$$
$$\text{Difference.} \quad 10 = +8 + 2$$

We then, as above, take away from the right side -2 (i. e., negative units) and hence must add $+2$ to the other side in order that the truth of the equation shall be maintained. In like manner we have :

$$- 8 = -8 + 2 - 2$$
$$\text{Subtracting} \quad + 2 = \quad +2$$
$$-10 = -8 - 2$$

And :

$$-2 = -2 - 8 + 8$$
$$\text{Subtracting} \quad -8 = \quad -8$$
$$+6 = -2 + 8$$

(3) Multiplication. The rule for the influence of the in the multiplication of opposed quantities is readily d from the conditions.

If both factors have the same sign the product is alw itive, but if the factors have different signs the pr negative. In other words : *Like signs give plus and unlike signs give minus.*

Here we have to consider what the multiplication by a negative quantity really means.

The sign prefixed to the multiplicand signifies the *sense* (or we may say, the *direction*), in which the *multiplicand* is to be taken. If we take a positive quantity in its opposite sense, it becomes negative ; as for example, $+8$ taken once in the opposite sense, becomes -8 ; $+8$ taken twice in the opposite sense becomes -16, etc. Therefore : $-1 \cdot +8 = -8$; $-2 \cdot +8 = -16$, etc. But if we take a negative quantity again in the same sense, the second negative overcomes the first * and gives a positive result. Thus : -8 taken once in the *same* sense, gives $+8$; or taken twice in the same sense gives $+16$, etc., or $-1 \cdot -8 = +8$; $-2 \cdot -8 = +16$, etc.

Now, if we take a negative quantity, once, just as it is, i. e., multiply it by $+1$, we get the quantity unchanged. That is, $+1 \cdot -8 = -8$; or if we take it twice we get -16, etc. The same is true of $+1 \cdot +8 = +8$; $+2 \cdot +8 = +16$, etc. We append a few examples.

Multiplicand	$+9$	-9	$+9$	-9
Multiplier	$+3$	$+3$	-3	-3
Product	$+27$	-27	-27	$+27$

We have for the general definition of multiplication, applicable alike to whole numbers or fractions, whether positive or negative, the following : *Multiplication consists in treating one quantity, the multiplicand, exactly the same as unity is treated by the other quantity, the multiplier.†*

For example, in $\frac{3}{4} \cdot \frac{5}{7}$ we have in the case of the multiplier $\frac{3}{4}$, the fourth part of unity taken three times, therefore, we must in the multiplication have the fourth part of $\frac{5}{7}$ taken three times : that is we have $\frac{5}{28}$ taken three times, and $\frac{3}{4} \cdot \frac{5}{7} = 3 \cdot \frac{5}{28} = \frac{15}{28}$.

In the product $-3 \cdot 8$ we have unity taken three times in the negative sense, and therefore we must have also 8 taken three

---

*Just as in English two negatives make an affirmative. "I have not nothing $=$ I have something."

†This definition is given in Thibaut's Arithmetic, and in Cauchy's " *Cours d'Analyse.*" It does not apply to irrational or imaginary quantities."

times in the negative sense, or $-3 \cdot 8 = -8-8-8 = -24$. In the same way $-3 \cdot -8 = +8+8+8 = 32$.

(4) Division. The rule for the division of opposed quantities follow from the rule for multiplication, namely : if the dividend and the divisor have the same signs, the quotient will be positive, but if they have different signs the quotient will be negative. In other words : *Like signs give plus, unlike signs give minus.* The quotient must always be a quantity which when multiplied by the divisor, will give the dividend. For example :

$$\frac{+8}{+2} = +4 \; ; \text{ because } +4 \cdot +2 = +8$$

$$\frac{+8}{-2} = -4 \; ; \text{ because } -4 \cdot -2 = +8$$

$$\frac{-8}{+2} = -4 \; ; \text{ because } -4 \cdot +2 = -8$$

$$\frac{-8}{-2} = +4 \; ; \text{ because } +4 \cdot -2 = -8$$

321.

Note to §92. (1) From the multiplication of two polynomial quantities we obtain a relation which may often be used to simplify the division of polynomials. If, for example, we multiply $5ac - \frac{2}{3}bc$ by $7ax + \frac{4}{5}bx$, we have :

$$\left. \begin{array}{l} 5ac - \frac{2}{3}bc \\ 7ax + \frac{4}{5}bx \end{array} \right\} \text{factors}$$

$$35a^2cx - \frac{14}{3}abcx$$
$$+ 4\,abcx - \frac{8}{15}b^2cx$$
$$\overline{35a^2cx - \frac{2}{3}\,abcx - \frac{8}{15}b^2cx}$$

Now it is evident that this product is divisible by either of the two factors by the multiplication of which it has been produced.

Let us now take the reverse case, and assuming the result to be unknown, divide $35a^2cx - \frac{2}{3}abcx - \frac{8}{15}b^2cx$ by $5ac - \frac{2}{3}bc$.

It is clear that the quotient will contain more than one member, because the dividend contains more members than the divisor ; also that one member of the quotient must be a quantity which, when multiplied by the whole divisor, will at least give one member of the dividend. From these considerations we deduce the following rule : Divide one member of the dividend (say $35a^2cx$) by one member of the divisor, (say $5ac$) and put the quotient $\left(\dfrac{35a^2cx}{5ac}=7ax\right)$ as *one part* of the desired quotient ; then multiply this by the entire divisor and subtract the product from the dividend. The same operation repeated upon the remaining portion of the dividend will give the second number of the quotient, etc., as the following example will show :

$$5ac-\tfrac{2}{3}bc \;\big|\; \begin{array}{l} 35a^2cx-\tfrac{2}{3}\,abcx-\tfrac{8}{15}b^2cx \\ 35a^2cx-1\tfrac{1}{3}\,abcx \\ \hline \quad\; 4\,abcx-\tfrac{8}{15}b^2cx \\ \quad\; 4\,abcx-\tfrac{8}{15}b^2cx \end{array} \;\big|\; 7ax+\tfrac{4}{5}bx$$

It is clear that if the first time we divide $35a^2cx$ by $5ax$, that the quotient $=\dfrac{35a^2cx}{5ac}=7ax.$

Also $5ac$ is evidently contained in the first number of the remainder $\dfrac{4abcx}{5ac}=\dfrac{4}{5}bx$ times.

If, instead of dividing the first member of the dividend by $5ac$, we had divided the second or third member, the resulting quotient would have been a fraction, which would not have been so convenient. It therefore follows that the arrangement of the members of the dividend is not a matter of indifference. The members of both dividend and divisor should be so arranged that they stand in the orders of the ascending or descending powers of one and the same symbol. The reason for this arrangement will appear if we multiply together two factors which are so arranged. In the preceding example, both dividend and divisor were arranged according to the descending powers of a, namely, a^2, a^1, a^0, in the dividend, and a^1, a^0, in the divisor, and also according to the descending powers of b. i. e., b^2, b^1, b^0.

For practice work out the following indicated divisions :

$$\frac{a^2-b^2-ac+bc}{a-b}=a+b-c$$

$$\frac{x^4-x^3-7x^2+x+6}{x^2-x-6}=x^2-1$$

$$\frac{x^4-x^3-7x^2+x+6}{x+1}=x^3-2x^2-5x+6$$

$$\frac{x^2+ax+b}{x+a}=x+\frac{b}{x+a}$$

(2) According to §§ 254, 256, in the following geometrical series :

$$a^{n-1}+a^{n-2}x+a^{n-3}x^2+a^{r-4}x^3+\ldots+x^{r-1}$$

in which a^{n-1} is the first term, x^{n-1} the last term, and $\dfrac{x}{a}$ the exponent, the sum will be $\dfrac{x^n-a^n}{x-a}$. This gives us directly the law : *Any expression of the form x^n-a^n must be divisible by $x-a$ without a remainder, n being any whole number.* Thus we have directly as a result of the above rule :

$$\frac{a^3-b^3}{a-b}=a^2+ab+b^2 ;$$

$$\frac{a^4-b^4}{a-b}=a^3+a^2b+ab^2+b^3 ;$$

$$\frac{x^5-1}{x-1}=x^4+x^3+x^2+x+1.$$

(3) The expression a^n-b^n is also divisible by $a+b$, without a remainder, when n is an *even* number, but not otherwise. The expression a^n+b^n is also divisible by $a+b$ without remainder when n is an odd number, but not otherwise.

322.

Note concerning Proportions, etc. (1) When two quantities, a and b, have the same ratio to each other as two other quantities, c and d, so that a divided by b gives the same quotient as c divided by d, we can form from these two factors of equal value the following equation :

$$\frac{a}{b}=\frac{c}{d}$$

Such a equation is called in the older phraseology, a *Proportion*, and written, instead of as above :

$$a:b=c:d$$

This is read, a is to b, so is c to d ; that is, a is divided by b as often as c is by d. The quantity a is called the first term, and the quantities b, c, and d, the second, third and fourth terms, respectively ; also a and d are called the *extremes*, and b and c the mean terms, or briefly, the *means*. Thus we have for the four quantities 2, 4 ; 3, 6 the following proportion :

$$2:4=3:6, \text{ or } \tfrac{2}{4}=\tfrac{3}{6}.$$

If the two middle terms are alike, as $a:b=b:d$ the proportion is a *continual* one, and the quantity b is said to be a *mean proportional* to the other two.

Thus we have 3 is to 6 as 6 is to 12, or

$$3 : 6 = 6 : 12$$

and 6 is a mean proportional between 3 and 12.

If the four terms of a proportion are undenominate numbers, or are all of the same denomination we have the following relations ; applications of which will be found in Geometry.

If : $a : b = c : d$, or $3 : 6 = 2 : 4$

we have also : $ad = bc$, or $3 \cdot 4 = 6 \cdot 2$

Proof. From $\dfrac{a}{b} = \dfrac{c}{d}$, we have, by multiplying both sides by bd,

$$ad = bc$$

If one term in a proportion is unknown, it can easily be determined.

If, for example, it is required to find x in the following proportion :

$$3 : 5 = 9 : x$$

we have from the above proof, the product of the extremes is equal to the product of the means, or :

$$3x = 5 \cdot 9$$

whence : $x = \frac{5 \cdot 9}{3} = 15$

Likewise, if :

$$3 : x = 9 : 15, \ 9x = 45, \text{ and } x = 5$$
$$x : 5 = 9 : 15, \ 15x = 45, \text{ and } x = 3$$
$$a : x = b : c, \ bx = ac, \text{ and } x = \frac{ac}{b}$$

In this way, also, we find the mean proportional between two given quantities is equal to the square root of their product. Thus, if we are required to find the mean proportional between the following quantities : 3 and 12 ; 5 and 2 ; a and b ; we have:

$$3 : x = x : 12, \text{ whence } x^2 = 36, \text{ and } x = \pm 6$$
$$5 : x = x : 2, \text{ whence } x^2 = 10 ; \ x = \sqrt{10} = \pm 3.162 \ldots$$
$$a : x = x : b, \text{ whence } x^2 = ab, \text{ and } x = \sqrt{ab}$$

The problem to divide a quantity a, in *continuous* proportion means to divide a into two such parts that the smaller part shall be to the larger part, as the larger part is to the whole quantity.

(2) In every proportion the relation of the terms can be reversed. Thus, in the proportions

$$a : b = c : d \; ; \quad \frac{a}{b} = \frac{c}{d} \; ; \quad 15 : 3 = 20 : 4$$

we can write also :

$$b : a = d : c \; ; \quad \frac{b}{a} = \frac{d}{c} \; ; \quad 3 : 15 = 4 : 20$$

(3) In every proportion the relation of the first term to the third is the same as that of the second to the fourth. Thus, in the proportions :

$$a : b = c : d \; ; \quad \frac{a}{b} = \frac{c}{d} \; ; \quad 3 : 6 = 9 : 18$$

we have also : $a : c = b : d \; ; \quad \dfrac{a}{c} = \dfrac{b}{d} \; ; \quad 3 : 9 = 6 : 18$

(4) In every proportion, the relation of the sum or difference of the first two terms is to either the first or second term, as the sum or difference of the last two terms is to the third or fourth term. Expressed in symbols :

$$a : b = c : d \qquad\qquad \text{or} \quad 8 : 2 = 12 : 3$$
$$a \pm b : a = c \pm d : c \; ; \qquad 10 : 2 = 15 : 3$$
$$\text{or} \quad a \pm b : b = c \pm d : d \; ; \qquad \text{or} \quad 6 : 2 = 9 : 13$$

Proof.   We have from :

$$\frac{a}{b} = \frac{c}{d}$$

$$\frac{a}{b} \pm 1 = \frac{c}{d} \pm 1 \quad \text{or} \quad \frac{a \pm b}{b} = \frac{c \pm d}{d}$$

Also : $\dfrac{b}{a} \pm 1 = \dfrac{d}{c} \pm 1 \quad \text{or} \quad \dfrac{b \pm a}{a} = \dfrac{d \pm c}{c}$

$$\text{or} \quad \frac{a \pm b}{a} = \frac{c \pm d}{c}$$

323.

When several ratios are equal to each other, the sum of all the first terms will be to the sum of all the second term as any first term is to its second term. Expressed in symbols, we have, if :

$$A : a = B : b = C : c = D : d = E : e, \text{ etc.,}$$

then will :

$$A + B + C + D + E \ldots : a + b + c + d + e \ldots = A : a = B : b, \text{ etc.}$$

Proof. Call the common value of all the ratios $=e$, and we have from the first series:

$$\frac{A}{a}=e\,; \quad A=ae\,; \qquad\qquad \frac{B}{b}=e\,; \quad B=be\,;$$

$$\frac{C}{c}=e\,; \quad C=ce\,; \qquad\qquad \frac{D}{d}=e\,; \quad D=d\,;$$

$$A+B+C+D+\ldots=ae+be=ce+de+\ldots$$

$$A+B+C+D+\ldots=(a+b+c+d+\ldots)e$$

$$\frac{A+B+C+D+\ldots}{a+b+c+d+\ldots}=e=\frac{A}{a}=\frac{B}{b}=\text{etc.}$$

If, therefore, several fractions are equal to each other, then also each fraction is equal to the fraction of which the numerator and denominator are equal to the sums of all their numerators and denominators, thus:

$$\tfrac{2}{3}=\tfrac{8}{12}=\tfrac{4}{6}=\tfrac{14}{21}=\tfrac{6}{9}=\tfrac{2+8+4+14+6}{3+12+6+21+9}=\tfrac{34}{51}$$

324.

Note to § 214. If it is required to extract the square root of a polynomial expression which is in the form of a complete square, we must follow the same rule as with numbers, i. e., proceed according to the formula:

$$a^2+2ab+b^2\,; \text{ see § 191.}$$

Thus, for example:

$$\sqrt{(4x^4-12ax^3+29a^2x^2-30a^3x+25a^4}=\pm(2x^2-3ax+5a^2)$$

$$a^2=4x^4$$

$$\overset{2a}{4x^2},\ \overset{b}{-3ax}\,|{-12ax^3+29a^2x^2}$$

$$2ab+b^2=\ -12ax^3+9a^2x^2$$

$$\overset{2a}{4x^2}-6ax,\ \overset{b}{5a^2}|20a^2x^2-30a^3+25a^4$$

$$2ab+b^2=20a^2x^2-30a^3x+9ax^4$$

A similar principle holds good for cube root.

325.

Note to § 216. Magnitudes such as $\sqrt{-4}$; $\sqrt{-a}$; $\sqrt[4]{-a}$, in which the root symbol with an *even* exponent stands before a negative quantity, are called by some mathematicians *impossible* quantities, but by others they are termed *imaginary* quantities; for although no root can be extracted in numerical terms, yet such expressions can be used in connection with the so-called *real* quantities (that is those which can be numerically expressed with more or less degree of exactness), and many calculations which would otherwise be very tedious and difficult, be thus greatly simplified. The term *impossible* is certainly incorrect when applied to such quantities, for they are certainly *not* impossible in the sense in which we would regard, for example, such expressions, as a triangular circle, or a circular triangle, etc. The name *imaginary quantities*, is, however, not much better. It seems to imply something which can be formed in the imagination, but not in reality, such as "castles in the air." But it will be found that the class of magnitudes of which we are speaking, can exist not only in the imagination, but also in reality, hence the unsuitability of this name. The word "imaginary" in fact, is applied properly, not to the quantity itself, but only to the *numerical extraction of the root*, which latter has no real existence, so that we can say correctly that the numerical value of it *is* imaginary.

If this peculiar class of magnitudes is to be named, the title suggested by Gauss is far better, namely, *lateral quantities*, since this conveys a definite meaning.*

For the present we must consider these expressions as if they were actual results, that is, as symbolical quantities, just as any other symbol, and we can then apply to them without difficulty the same rules and methods as to the so-called real quantities.

The calculations with lateral quantities may often be simplified by separating the expression under the radical sign into two factors, one factor being the quantity itself with reversed sign, and the other factor being -1, and then indicating the root of each factor separately, § 211.

Thus we have, for example:

$$-4 = 4(-1), \text{ whence:}$$
$$\sqrt{-4} = \sqrt{4(-1)} = 2\sqrt{-1}$$
$$\sqrt{-5} = \sqrt{5(-1)} = \sqrt{5}\sqrt{-1}$$
$$\sqrt{-9} = \sqrt{9(-1)} = 3\sqrt{-1}$$
$$\sqrt{-a} = \sqrt{a(-1)} = \sqrt{a}\sqrt{-1}$$

* This matter will be more fully discussed in Lubsen's Analysis

Each lateral quantity may therefore be considered as the product of two factors, of which one is a real quantity, and the other is $\sqrt{-1}$, which latter we may consider as a special kind of unit, and to which we must apply the same methods and rules as to any other symbolical unit. Gauss suggested that the expression $\sqrt{-1}$ be represented more briefly by "i," and hence $\pm\sqrt{-1}$ by $\pm i$.

We must consider also the difference between the even and the odd powers of $\sqrt{-1}$, as these are by no means alike. We have :

$$(\sqrt{-1})^1=\sqrt{-1}$$
$$(\sqrt{-1})^2=\sqrt{(-1)(-1)}=\sqrt{(-1)^2}=-1 \quad (\S\, 216)$$
$$\text{or,} \quad (\sqrt{-1})^2=[(-1)^{\frac{1}{2}}]^2=(-1)^1=-1$$
$$(\sqrt{-1})^3=(\sqrt{-1})^2\cdot\sqrt{-1}=(-1)\sqrt{-1}=-\sqrt{-1}$$
$$(\sqrt{-1})^4=(\sqrt{-1})^2(\sqrt{-1})^2=(-1)(-1)=1$$
$$\text{or,} \quad (\sqrt{-1})^4=[(-1)^{\frac{1}{2}}]^4=(-1)^2=1$$
$$(\sqrt{-1})^5=(\sqrt{-1})^4\cdot\sqrt{-1}=1\sqrt{-1}=\sqrt{-1}$$

We see, therefore, that the first power of $\sqrt{-1}=\sqrt{-1}$, the second power $=-1$, the third power $=-\sqrt{-1}$, and the fourth power $=1$, beyond which these expressions repeat themselves. If, therefore, we let n be any whole number other than 0 or 1, we have in general :

$$(\sqrt{-1})^{4n}=1, \quad \text{or} \quad i^{4n}=1$$
$$(\sqrt{-1})^{4n+2}=-1, \quad \text{or} \quad i^{4n+2}=-1$$
$$(\sqrt{-1})^{4n+1}=\sqrt{-1}, \quad \text{or} \quad i^{4n+1}=i$$
$$(\sqrt{-1})^{4n+3}=-\sqrt{-1}, \quad \text{or} \quad i^{4n+3}=-i$$

326.

EXAMPLES.

(1) $\quad \sqrt{-1}+3\sqrt{-1}+\sqrt{-1}=5\sqrt{-1}$
$\sqrt{-4}+\sqrt{-9}=2\sqrt{-1}+3\sqrt{-1}=5\sqrt{-1}$
$\sqrt{-a}+\sqrt{-b}-\sqrt{-c}=(\sqrt{a}+\sqrt{b}+\sqrt{c})\sqrt{-1}$
$a\sqrt{-4}-\sqrt{-a^2}=2a\sqrt{-1}-a\sqrt{-1}=a\sqrt{-1}$

(2) $\quad \sqrt{-a}\cdot\sqrt{-b}=\sqrt{a}\sqrt{-1}\cdot\sqrt{b}\sqrt{-1}=-\sqrt{ab} \quad (\S\,216)$
$\sqrt{-3}\cdot\sqrt{-12}=-\sqrt{36}=-6 \quad (\S\,216)$

(a) Each of the combined real and lateral expressions may be called a "complex" quantity. An example of such a "complex" quantity is $2+3\sqrt{-1}$. If we represent the real portion by a, and the factor or coefficient of the lateral quantity by b, every complex expression may be reduced to the form: $a+b\sqrt{-1}$, or $a+bi$.

EXAMPLES.

(3) $4+6\sqrt{-1}+\sqrt{-16}+2=6+10\sqrt{-1}$

$3+3\sqrt{-9}-3-2\sqrt{-4}=0+5\sqrt{1}=5\sqrt{-1}$

$a+b\sqrt{-1}+a-b\sqrt{-1}=2a$

(4) $(a+b\sqrt{-1})(a-b\sqrt{-1})=a^2+b^2$ §91.

$(a+b\sqrt{-1})^2=a^2-b^2+2ab\sqrt{-1}$ §186.

$(a-b\sqrt{-1})^2=a^2-b^2-2ab\sqrt{-1}$

$(x+a+b\sqrt{-1})(x+a-b\sqrt{-1})=(x+a)^2+b^2$

(b) It will be shown in analysis, that as many different roots can be extracted from positive or negative unity, or from any given quantity, as there are units in the exponent.

Thus, for example, there are three different quantities which raised to the 3d power will equal 8 ; four different quantities which, raised to the fourth power will equal 1, etc. This we will only mention here, giving also the following examples :

$$(+2)^3=8$$
$$(-1+\sqrt{-3})^3=(-1)^3+3(-1)^2\sqrt{-3}+3(-1)(\sqrt{-3})^2+(\sqrt{-3})^3=$$
$$=1-3\sqrt{-3}+9-3\sqrt{-3}-3=8 \qquad (\S 213,2)$$
$$(-1-\sqrt{-3})^3=+1-3\sqrt{-3}+3\sqrt{-3}+9=8$$

whence : $\sqrt[3]{8}=2$; $=-1+\sqrt{-3}$; $=-1-\sqrt{-3}$

$1^4=1$; $(-1)^4=1$; $(\sqrt{-1})^4=1$; $(-\sqrt{-1})^4=1$

whence : $\sqrt[4]{1}=1$; $=-1$; $=-\sqrt{+1}$.

(5) $\dfrac{\sqrt{-1}}{\sqrt{-1}}=1;$ $\dfrac{6\sqrt{-1}}{2\sqrt{-1}}=3;$ $\dfrac{-\sqrt{-1}}{+\sqrt{-1}}=-1$

$\dfrac{\sqrt{-9}}{\sqrt{-4}}=\dfrac{3\sqrt{-1}}{2\sqrt{-1}}=1\tfrac{1}{2};$ $\dfrac{a\sqrt{-1}}{b\sqrt{-1}}=\dfrac{a}{b}$

$\dfrac{1}{\sqrt{-1}}=\dfrac{\sqrt{-1}}{\sqrt{-1}\cdot\sqrt{-1}}=\dfrac{\sqrt{-1}}{-1}=-\sqrt{-1}$

$\dfrac{a}{a+\sqrt{-b}}=\dfrac{a(a-\sqrt{-b})}{(a+\sqrt{-b})(a-\sqrt{-b})}=\dfrac{a^2-a\sqrt{b}\cdot\sqrt{-1}}{a^2+b}$

$$\frac{a+b\sqrt{-1}}{a-b\sqrt{-1}}=\frac{(a+b\sqrt{-1})(a+b\sqrt{-1})}{(a-b\sqrt{-1})(a+b\sqrt{-1})}=\frac{a^2-b^2+2ab\sqrt{-1}}{a^2+b^2}$$

327.

When it is required to square a binomial *numerical* expression of the form $\sqrt{a}\pm\sqrt{b}$ it is evident that we shall obtain, in general, a binomial expression of the form $A\pm\sqrt{B}$, consisting of a rational and an irrational part. Thus, for example :

$$(\sqrt{2}+\sqrt{3})^2=5+2\sqrt{6}=5+\sqrt{24}$$
$$(\sqrt{5}-\sqrt{3})^2=8-2\sqrt{15}=8-\sqrt{60}$$

Conversely, the square root of a numerical expression of the form $A\pm\sqrt{B}$, will in general be an expression of the form $\sqrt{a}\pm\sqrt{b}$, in which, according to circumstances, one of the parts may be rational. The rule by which these roots may be found, is readily deduced. We have :

$$\sqrt{(5+\sqrt{24})}=\sqrt{x}+\sqrt{y}$$

whence :

$$5+\sqrt{24}=x+y+2\sqrt{xy}$$

Now, if we seek such values of x and y as shall make the rational and irrational members of each side of the equation equal to each other we shall have :

(1) $\qquad\qquad x+y=5$

(2) $\qquad\qquad 2\sqrt{xy}=\sqrt{24}$

whence : $\qquad x^2+2xy+y^2=25$, and $4xy=24$

Subtracting we get :

$$x^2-2xy+y^2=1, \text{ whence :}$$

(3) $\qquad\qquad x-y=\pm 1$

From (1) and (3) we have $x=3$, $y=2$, and hence :

$$\sqrt{(5+\sqrt{24})}=\sqrt{3}+\sqrt{2}$$

In order to find $\sqrt{(\pm a\pm\sqrt{-b})}$, or, for example, $\sqrt{(-3+\sqrt{-16})}$, we proceed as follows :

Let : $\qquad \sqrt{(-3+\sqrt{-16})}=\sqrt{x}+\sqrt{-y}$

whence : $\qquad -3+\sqrt{-16}=x-y+2\sqrt{-xy}$

(1) $x-y=-3$ $\qquad\qquad x^2-2xy+y^2=9$

(2) $2\sqrt{-xy}=\sqrt{-16}$ $\qquad\qquad 4xy=16$

$$x^2+2xy+y^2=25$$

(3) $\quad x+y=\pm 5$

From (1) and (3) we have : $x=1$; $y=4$, whence :

$$\sqrt{(-3+\sqrt{-16})}=1+2\sqrt{-1}$$

328.

In some especial cases fractional algebraic expressions, when the numerical values are substituted for their symbols, reduce to the expression $\frac{0}{0}$, which we must be careful not to confound with o, or to consider as meaningless.

In order to show that $\frac{0}{0}$ may be a real expression, and that, in general, for each special substitution it has a special meaning, let us multiply the numerator and denominator of the algebraic expression :

$$\frac{b(a+b)}{a}$$

by $a-b$, by which we do not alter its value. We then have :

$$\frac{b(a+b)}{a} = \frac{b(a+b)(a-b)}{a(a-b)}$$

or :
$$\frac{b(a+b)}{a} = \frac{b(a^2-b^2)}{a(a-b)}$$

Both of these last expressions must give the same value when any numerical values are substituted for the letters. If we let $a=4$, $b=2$, we have $\frac{b(a+b)}{a}=3$, and $\frac{b(a^2+b^2)}{a(a-b)}=3$. But if we put $a=4$, $b=4$, we get : $\frac{b(a+b)}{a}=8$, and $\frac{b(a^2-b^2)}{a(a-b)}=\frac{0}{0}=\frac{0}{0}$.

Again, if we take $a=5$, $b=5$, we get for the first expression $=10$, and for the second $=\frac{0}{0}$.

We see therefore that for certain relations between a and b (in this case when $a=b$) that the fraction $\frac{b(a^2-b^2)}{a(a-b)}$ gives the indeterminate expression $\frac{0}{0}$, and also that the value of this expression depends upon the values of a and b. Thus if $a=5$, $b=5$, then $\frac{0}{0}=10$; if $a=3$, $b=3$, then $\frac{0}{0}=6$, etc.

In the above instance we see that we can readily avoid the introduction of the cause which makes the expression $\frac{b(a^2-b^2)}{a(a-b)}$ reduce to $\frac{0}{0}$, by reducing it to the form $\frac{b(a+b)}{a}$. Such a reduction is, however, not always possible, and we must then consider the expression $\frac{0}{0}$ as indeterminate, or else determine its value in some other way.

In the higher mathematics the expression $\frac{0}{0}$ often appears, but no regular rule has been deduced by which its value can be determined. In elementary algebra, therefore, it is sufficient to remember that the expression $\frac{0}{0}$ is not determinate, but may have at one time one value and at another time a different value.

The formula for the summation of a geometrical series is:

$$s = \frac{a\,(e^{n}-1)}{e-1}$$

In the case when $e=1$ (and the series therefore $=a+a+a+\ldots$) the sum evidently becomes na, while the value obtained from the formula is:

$$s = \frac{a(1^{n}-1)}{1-1} = \frac{0}{0} = na$$

The expression $\frac{\sqrt{(8+a)}-3}{a^{2}-1}$ becomes $=\frac{0}{0}$ when $a=1$, but it is also easy to see that for $a=1$, $\frac{0}{0}=\frac{1}{12}$, if we multiply numerator and denominator by $\sqrt{(8+a)}+3$, thus:

$$\frac{[\sqrt{(8+a)}-3]\,[\sqrt{(8+a)}+3]}{(a^{2}-1)\,[\sqrt{(8+a)}+3]} = \frac{a-1}{(a^{2}-1)\,[\sqrt{(8+a)}+3]}$$

whence: $\quad \dfrac{\sqrt{(8+a)}-3}{a^{2}-1} = \dfrac{1}{(a+1)\,[(8+a)+3]}$

329.

There exist in nature many mathematical researches which, we cannot fail to realize, extend into the realm of the infinite. The capacity of the mind, as well as the body, has its limits, and it must be admitted that there is a multitude of phenomena which extend beyond these limits. For example, no man can consider time, or space, as other than infinite (i. e., endless). Although all mathematical investigations in which the relations of the infinite are concerned, belong properly to the so-called analysis of the infinite, yet there are among them some cases which permit elementary discussion, and since these cases appear in some branches of applied mathematics (such as Mechanics and Geometry), we shall here briefly glance at them. We must here ask the reader's indulgence and assistance in our attempts to illustrate this subject; for that which belongs to the mind cannot be grasped by the hands, and in discussing such subjects it is easy to argue in unending circles and make the matter darker than ever, by futile attempts at explanation.

Let us first consider the following geometrical progression :

$$\tfrac{1}{2}+\tfrac{1}{4}+\tfrac{1}{8}+\tfrac{1}{16}\cdots\tfrac{1}{2^n}.$$

It is evident that in this series the terms become continually smaller and smaller, and that n can be taken so large that $\dfrac{1}{2^n}$ will be smaller than we can describe, or even conceive.

If we sum up, according to § 254, some of the first members of the series we see that the value will be almost exactly the same for a billion terms as for a million terms. In both cases the sum is almost equal to unity. No matter how many terms we give the series, its sum can never equal unity, and in no case can it possibly be greater than unity.

These facts, perhaps curious to the beginner, we can readily deduce in the following manner. For any number of terms n, no matter how great the number, we have $\dfrac{1}{2^n}$ for the last term, and hence can apply the general formula for the sum :

$$s=\frac{le-a}{e-1}$$

in which for the case under consideration $a=\tfrac{1}{2}$; $e=\tfrac{1}{2}$; $l=\dfrac{1}{2^n}$.

Substituting these values we have :

$$s=\frac{\dfrac{1}{2^n}\cdot\dfrac{1}{2}-\dfrac{1}{2}}{\tfrac{1}{2}-1}=\frac{\dfrac{1}{2}-\dfrac{1}{2}\cdot\dfrac{1}{2^n}}{1-\tfrac{1}{2}} \qquad (\S\ 254)$$

from which, multiplying the numerator and denominator by 2, we have :

$$s=1-\frac{1}{2^n}.$$

We therefore see that s can never be greater than 1, but can be brought so nearly equal to unity that the difference shall be smaller than any assignable quantity ; for the greater n becomes, the smaller will $\dfrac{1}{2^n}$ be.

We now ask : how far must the series $\tfrac{1}{2}+\tfrac{1}{4}+\ldots$ be carried, or how great must n be conceived, in order that this difference $\dfrac{1}{2^n}$ shall become so small that we shall be unable to assign any smaller value to it ?

Answer. No matter how small a magnitude may be conceived, yet since any actual quantity must be conceivable, we can conceive a smaller one. In order to make $\frac{1}{2^n}$ so small that no smaller value can be assigned to it, brings us to the *infinitely* small. This, then, is a real "non-plus-ultra," a "thus far and no farther," for if there were any "beyond," there could be no infinity. We are therefore compelled to admit that n must be infinitely great, (represented by ∞), whence $\frac{1}{2\infty} = \frac{1}{\infty}$, and consider the series as practically endless. So long as the terms of the series can be conceived as growing smaller and smaller, so long the series itself (that is, the number of its terms) cannot be actually considered as infinite. The idea of an infinite quantity is in every case a mental conception which is itself without limit, forming a thought to which no limits can be set, and this it is which is sought to be conveyed by the word *infinite*.

It is therefore evident that if we for a moment consider the number of terms of the series: $\frac{1}{2} + \frac{1}{4} + \frac{1}{8} \dots \frac{1}{2^n}$, as infinitely great, there is bound up with this conception also the conception that the last term must be infinitely small; that is, so small that it cannot be conceived any smaller.

Again, if we conceive n as infinitely great, we must necessarily also conceive the series as extending to infinity, and therefore actually not divisible so that we could take, for instance, half of its terms. Against this latter view, the ordinary reasoning will, however, be apt to rebel. This objection will disappear, however, if we do not consider ∞ as a number, that is, a collection of units, but rather as the *absence of number*—a quantity so great that it cannot be made any greater. If we hold fast to this point, that infinity cannot possibly be exceeded or increased, we must see that the terms of this infinite series, $\frac{1}{2} + \frac{1}{4} + \dots$ not only approximately, but actually, must reach o in value. For its terms must extend until they become smaller than any value which it is possible to assign. But a quantity so small that no value can be assigned to it, and which cannot be further subdivided, is not a real magnitude at all. In order, however, not to cut the thread of our investigation or lose connection with the argument, such a constantly diminishing value is considered as always being a real value, and in order to be able to use it in calculations it is expressed by $\frac{1}{\infty}$, by which is meant an *infinitely small* quantity (an indivisible magnitude, a differential, fluxion, etc.; the beginning or ending, without actual magnitude itself, but which have been, or will be, a real magnitude).

Concerning the form $\frac{1}{\infty}(dx)$, we shall study by and by. The subject cannot be pursued here further, as it belongs to the Infinitesimal Calculus (Differential and Integral), which forms in itself a branch of higher mathematics, requiring a separate treatise.*

We shall here show by examples, however, that the members of a constantly diminishing geometrical series become ultimately really, not approximately, indivisible, and that hence the infinitely small quantity $\frac{1}{\infty}$ is really a magnitude, which, being no longer capable of subdivision, is properly placed equal to zero.

(1) Suppose we consider the lifetime of a man as unity. Then suppose $\frac{1}{2}$ of this to elapse, then let one-half of the remaining half, or $\frac{1}{4}$ of the whole, pass, and so on to infinity, we have $\frac{1}{2}+\frac{1}{4}+\frac{1}{8}+ \ldots +\frac{1}{\infty}$. But if the man's life is to end, we must consider the last member of the series as a mere instant, a moment, which cannot be sub-divided into smaller parts.

If it were not so, since time is a continuously passing quantity, there would still always be a small amount of time remaining, i. e , the man would live forever, which, besides being contrary to our experience of nature, is also contrary to our original premise that the life should terminate at a definite period of time.

All real quantities have the common property that each is composed of innumerable similar parts, and is therefore capable of being subdivided. If it were not so we could not conceive any unit or scale by which these quantities could be measured, neither could they be represented by numbers; in short we would be without the conception of *quantity*. If then the last *instant* of time, in the above illustration, cannot be further divided into any conceivable smaller periods of time, it is evident that such an indivisible or infinitely small quantity, $\frac{1}{\infty}$, if it cannot exist in the imagination, and cannot be used interchangeably with zero, at least cannot be used in comparison with real values.

Hence it is clear that the usual expression: *A quantity can become smaller than any assignable value*, is practically the same as

---

*See Lubsen, Infinitesimal Calculus.

saying that an infinitely small quantity is no longer capable of sub-division.

Any assignable amount of time must have duration, as well as beginning and ending. An *instant*, however, has no assignable duration ; its beginning and its ending coincide, and hence instantaneity is non-divisible. Notwithstanding the fact that, because of this indivisibility, an instant can have no magnitude, yet it is certainly different from an absolute nullity or zero. No man can conceive of the passing of an actual zero, but every one can conceive of the nondivisible and nonmeasurable instant.

The greater the number of terms summed up of the series $\frac{1}{2}+\frac{1}{4}+\frac{1}{8}+\ldots$, the smaller will be the difference by which the sum differs from unity. Since this difference becomes smaller as the number of terms increases, and as it becomes ultimately smaller than any assignable value, it is evident that 1 is not only the *limit* which the sum cannot exceed, but it is *actually* the true sum of the whole *infinite series*. For, we get $s=1-\frac{1}{2\infty}$, and since the member $\frac{1}{2\infty}$ is indivisible and may be considered $=0$, therefore, $s=1$. We see, therefore, that a thing may be considered infinite in one sense, and yet real in another. The above series is infinite, with regard to the number of its terms, while its sum has the very definite value of 1.

It may be noted that in this series each term is equal to the sum of all the following terms :

$$\frac{1}{2}=\frac{1}{4}+\frac{1}{8}+\frac{1}{16}\cdots\frac{1}{\infty}$$
$$\frac{1}{4}=\frac{1}{8}+\frac{1}{16}+\frac{1}{32}\cdots\frac{1}{\infty}$$

Beginners sometimes think they find an inconsistency in this fact, since because each term is twice as great as the next following term, which we have said may be considered equal to 0, should when doubled, give the next to the last term, etc., and hence all the terms may thus be proved equal to 0.

This reasoning involves the error of assuming that the last term of an infinite series can have any *assignable* value. Such a series may be compared to a circle, which is absolutely without end.

All such infinite series may be summed up. Thus we have :

$$\frac{1}{3}+\frac{1}{9}+\frac{1}{27}+\cdots\frac{1}{\infty}=\frac{1}{2}$$
$$\frac{1}{5}+\frac{1}{20}+\frac{1}{80}+\cdots\frac{1}{\infty}=\frac{4}{15}$$
$$\frac{1}{3}+\frac{1}{6}+\frac{1}{12}+\cdots\frac{1}{\infty}=\frac{2}{3}$$

(2) According to the rule in §318 we have for the value of the endless periodical decimal $0.3636\ldots = \frac{4}{11}$, namely:

$$s = 0.363636 \qquad (1)$$

$$100s = 36.363636 \qquad (2)$$

$$99s = 36$$

$$s = \frac{36}{99} = \frac{4}{11}$$

Although the correctness of this rule is undoubted, yet the objection has been made by some that the series of decimals in the second equation, on account of the moving of the decimal point, is not the same as that in equation (1), and that equation (2) should properly have two decimal places fewer than (1). This would make the difference between the two decimal portions not $= 0$. The argument is made that no matter if both decimals be extended to infinity, the one which started with two more decimals would always have two more. This, however, is a false conclusion. The difference between the finite cannot be measured by two steps. The infinite can neither be increased or diminished by the addition or subtraction of any finite quantity. Both series of decimals, when extended to infinity, are equal to each other ($\infty \pm a = \infty$).

The truth of the operation can be made evident by reversing the operation, thus:

$$s = \frac{4}{11} = 0.3636\ldots$$

$$100s = 36\frac{4}{11} = 36.3636\ldots$$

$$99s = 36 = 36.0000\ldots$$

(3) Looking at the matter in another light, we may consider the period of the fraction as a geometrical series, and proceed to find its sum. We may take the period as terms of such a series, in which $\frac{36}{100}$ is the first term, $\frac{1}{100}$ is the exponent, and the number of terms $= \infty$, so that the last term will be $\frac{1}{\infty} = 0$, we then have:

$$s = 0.3636\ldots = \frac{36}{100} + \frac{36}{10000} + \frac{36}{1000000} + \cdots \frac{1}{\infty}$$

whence:
$$s = \frac{\frac{1}{\infty}\,\frac{1}{100} - \frac{36}{100}}{\frac{1}{100} - 1} = \frac{0 - \frac{36}{100}}{-\frac{99}{100}} = \frac{36}{99} = \frac{4}{11}$$

330.

Under the title "Mathematical Sophisms," a volume of fallacies has been published at Vienna (1850) by Herr Viola; the book containing no explanation of the fallacies. We give here a pair of the most deceptive of these fallacies with explanations.

Herr Viola shows how one may prove that 4 is greater than 12. It is admitted that:

$$+7 > +5$$

now add: $\quad -8 = -8$

$$-1 > -3$$

then multiply: $\quad -4 = -4$

$$4 > 12$$

The assumption is here made is that because "equals added to equals, give equals," therefore equals added to unequals, give unequals, with the inequality on the same side of the sign.

This statement is by no means general, but is only true for the special conditions in which the addition or subtraction of equals to or from the inequality, gives the members of the inequality the same sign as before, (Example 1):

(1)	(2)	(3)
$+7 > +5$	$+7 > +5$	$+8 > +5$
$-4 = -4$	$-8 = -8$	$-6 = -6$
$+3 > +1$	$-1 < -3$	$+ \ldots -$

If the signs become changed the direction of the inequality sign must be reversed, (Example 2).

If the signs come out different on both sides (Example 3) no comparison of value is possible, and the sign of inequality becomes meaningless, because we cannot compare articles of a different kind, and can no more say that $+3$ is greater than -2 than we can say that 3 minutes are greater than 2 dollars.

The signs $+$ and $-$ are here "denominations" (§ 144). An *absolute* number has no sign whatever. A false idea, formerly too current, that a negative quantity was smaller than a positive quantity, or even smaller than 0, arose from the ideas of creditor and debtor, so that that the saying arose of a debtor that he owns less than nothing. The debtor might well wish his debts to be smaller than nothing, but mathematics does not deal with wishes. A quantity smaller than zero is impossible.

331.

The following fallacy was intended to demonstrate that all numbers are equal.

Suppose $a>b$, and also $a-b=c$, we then have :

$$(a-b)(a-b)=(a-b)c$$

or $\quad a^2-2ab+b^2=ac-bc \qquad (1)$

adding : $\quad -ac+ab-b^2=-ac+ab-b^2$

$$a^2-ab-ac=ab-b^2-bc \qquad (2)$$

or $\quad a(a-b-c)=b(a-b-c) \qquad (3)$

whence : $\qquad a=b$

Explanation. From the expression $a-b=c$, we have $ab-b^2=bc$. We have, therefore added to both sides of equation (1) the quantity $-ac+bc$, and equation (2) thereby becomes $o=o$. We also see that in equation (3) $a-b-c=o$, and the equation, therefore is really $a\cdot o=b\cdot o$.

But when o is a factor in any product the entire product becomes $=o$, notwithstanding the other factors, because zero cannot be increased by multiplication, neither can any arithmetical operation affect its value.

332.

In the following manner it was attempted to prove that five is equal to four.

Let $\qquad x+z=9$

then $\quad (x+z)(x-z)=9(x-z)$

and $\qquad x^2-z^2=9x-9z$

adding $\quad z^2-9x=z^2-9x$

$$x^2-9x=z^2-9z$$

completing squares by adding $\frac{81}{4}$, we have :

$$x^2-9x+\tfrac{81}{4}=z^2-9z+\tfrac{81}{4}$$

or $\qquad (x-\tfrac{9}{2})^2=(z-\tfrac{9}{2})^2 \qquad (1)$

$$x-\tfrac{9}{2}=z-\tfrac{9}{2} \qquad (2)$$

$$x=z$$

Equation (1) is false. This will be seen by substituting the values for x and z, (i. e., 4 and 5) which are really known, though ostensibly sought, we see that the root of the left side is positive $=\frac{1}{2}$ and the root of the right side is negative, $=-\frac{1}{2}$. Equation (2) should properly be written :

$$x-\tfrac{9}{2}=-(z-\tfrac{9}{2})$$

whence we simply get again :

$$x+z=9$$

The rule "equals treated equally give equals" is always true, but it by no means always follows that quantities which when treated alike give equals, are therefore in themselves equal.

We know that

$$(+a)^2 = (-a)^2$$

both sides being equal to $+a^2$, but it does not follow that

$$+a = -a$$

333.

We shall close with the following puzzle. There is an equation with one unknown quantity but with *no* root!

If we have:

$$1 + \sqrt{x-1} = +\sqrt{x-4}$$

we get, according to § 230: $x = 5$

But this value of x, when substituted in the equation, fails to satisfy it, although it is clearly derived from the equation. The reason for this puzzle lies in the fact that the equation is originally a false one, and no true conclusion can be drawn from false premises. The falsity of the equation can be seen by inspection, before beginning to attempt its solution. It is apparent that x cannot be any positive number greater than 4, for in such case:

$$\sqrt{x-1} > \sqrt{x-4}$$

Neither can x be greater than 4, or less than 1, or be negative, since a "lateral" or imaginary quantity cannot be equal to a real one. We see, for the same reasons, that x cannot be either a lateral or a complex quantity. How then can the value $x = 5$ be obtained from the equation?

We have just seen above that quantities which when treated alike give equals, are not always themselves equal. If we take for the *double* signs of the roots (§ 216) the lower or negative ones, we have a possible equation, namely:

$$1 - \sqrt{x-1} = -\sqrt{x-4}$$

which equation is satisfied by $x = 5$. It must be remembered that in all arithmetical operations the symbols of mathematics, like the letters of the alphabet, should never be used to promote thoughtless deceptions, but should be made to serve their true purpose of recording existing and possible knowledge.

INDEX.

CPSIA information can be obtained at www.ICGtesting.com
Printed in the USA
BVOW012256190213

313725BV00021B/629/P